STUDENT SOLUTIONS MANUAL TO ACCOMPANY

INTRODUCTORY STATISTICS

2nd Edition

Sheldon M. Ross

PREPARED BY

**LLOYD R. JAISINGH
MATHEMATICS & COMPUTER SCIENCE
DEPARTMENT
MOREHEAD STATE UNIVERSITY**

Elsevier Academic Press
30 Corporate Drive, Suite 400, Burlington, MA 01803, USA
525 B Street, Suite 1900, San Diego, California 92101-4495, USA
84 Theobald's Road, London WC1X 8RR, UK

ISBN 13: 978-0-12-088551-0
ISBN 10: 0-12-088551-4

For all information on all Elsevier Academic Press Publications
visit our Web site at www.books.elsevier.com

Working together to grow
libraries in developing countries

www.elsevier.com | www.bookaid.org | www.sabre.org

ELSEVIER   BOOK AID
            International   Sabre Foundation

Transferred to Digital Printing 2009

Contents   Page

Chapter 1 INTRODUCTION TO STATISTICS

INTRODUCTION TO STATISTICS

PROBLEMS

1. (a) Largest difference = 0.17; year 1946.

   (b) There were more years in which the average number of years of school completed by the older starting group exceeded that of the younger group; from 1947 to 1949 and from 1951 to 1952.

3. The graph below shows a plot of the percentages of U.S. adults, characterized by gender and educational level, that smoked in the years 1999 to 2002.

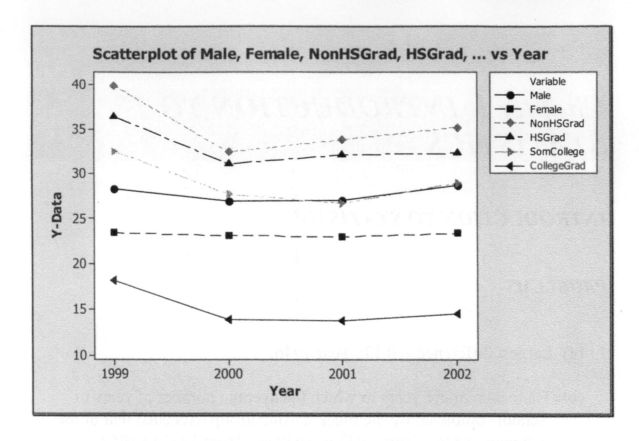

(a) The group for which there was a steady decline was the college graduates
   (CollegeGrad in graph). Note, however, a slight increase in 2002 for this group but not as severe as the rest of educational levels.

(b) Yes. A slight increase overall for all educational levels.

5. It is important that the researcher should not know which of the patients are receiving the new drug and which are receiving the placebo because the researcher should not be given an opportunity to involve his/her own bias in the sampling process. For example, researcher may subconsciously treat the "drug group" in a much friendlier manner and thus influencing the final outcome.

7. (a) The Literary Digest prediction for the 1936 election was so far off because only the rich or very rich would have been owners of automobiles and telephones at that time. Such a sample would be a

biased sample.

(b) The approach used by the Digest would work better today since a very large portion of the population now are owners of cars and telephones.

9. A reasonable conclusion that one can make is that the average death age for the deceased, as reported in the obituaries in the New York Times, is rather high. Since, more than likely, that most of the obituaries will likely be from well-known or affluent people from New York city and its vicinity, one may infer that this group of people seems to live longer than the general population.

11. (a) No. The answer could be no if the number of responses (86) is not large enough to be considered as a representative sample from the population of recent graduates. In addition, since $75,000 yearly salary is rather high for recent graduates, the university could be overestimating the average salary if most of the responses came from high paying occupations. In addition, if there are large outlying values in the data, this could also cause an overestimation.

Yes. The answer could be yes if the number of responses (86, 43% return) is large enough to be considered as a representative sample from the population of recent graduates and there are no large outlying salaries.

(b) If the answer is no, then if the sample was biased towards, for example, computer science and engineering, then maybe, these are high enough paying jobs which could have an annual average salary of $75,000.

13. Graunt's method of collecting data was not random and it was not a representative sample. He collected data from certain parishes (or neighborhoods) in London and then try to infer for the whole of London. He should have made inferences only on those parishes for which he collected data. In order for Graunt to make inferences on the total population of London, he probably should have collected data by

stratifying the London area with each parish as a stratum. His basic assumption was that all the parishes had the same death rate. That is, there was roughly one death for every 88/3 people for all the parishes in London.

15. One could use Graunt's information to help determine the average amount in annuities that would have to be paid out for the different age classifications. This in turn would help to determine the average cost

for

and the profits from, the annuities.

17. (a) Proportion of babies surviving to age 6 = (100 - 36)/100 = 0.64.

(b) Proportion of babies surviving to age 46

= {100 - (36 + 24 + 15 + 9 + 6)}/100 = 10/100 = 0.10.

(c) Proportion of babies dying between age 6 and age 36

= (24 + 15 + 9)/100 = 48/100 = 0.48.

Chapter 2   DESCRIBING DATA SETS

SECTION 2.2 FREQUENCY TABLES AND GRAPHS

PROBLEMS

1. (a) The frequency table is shown below where f represents the frequency for the different values of family sizes that reside in a small town in Guatemala.

Values	4	5	6	7	8	9	10	11	12	13	15
f	1	1	3	5	5	3	5	2	3	1	1

(b) Below is the line graph of *frequencies* versus *family size* for the data set in part (a).

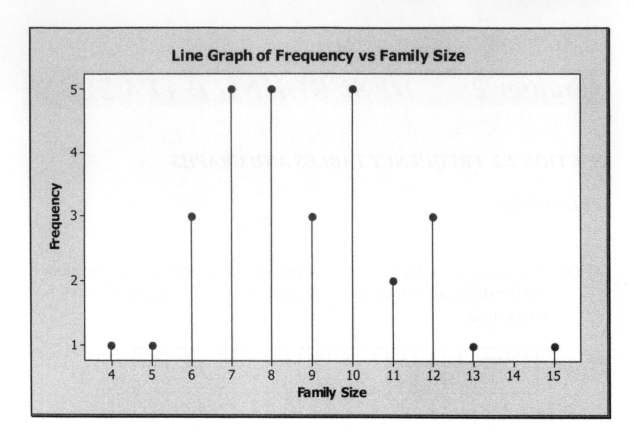

(c) Below is the frequency polygon graph of *frequencies* versus *family size* for the data set in part (a).

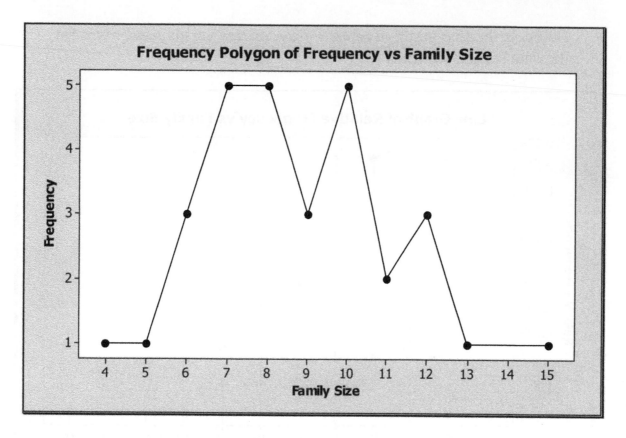

3. (a) 4; (b) 1; (c) 15; (d) 0.

5. The missing frequencies are in the shades cells. These frequencies will make the data set symmetric. In the frequency table, the frequency is denoted by *f*.

Value	10	20	30	40	50	60
f	8	3	7	7	3	8

7. The relative frequency table for Problem 1 is shown below. The relative frequency is denoted by *f/n*.

Value	4	5	6	7	8	9
f/n	.0333	.0333	.1000	.1667	.1667	.1000

Value	10	11	12	13	15	
f/n	.1667	.0667	.1000	.0333	.0333	

Below is the line graph of *relative frequencies* versus *family size* for the data set in Problem 1.

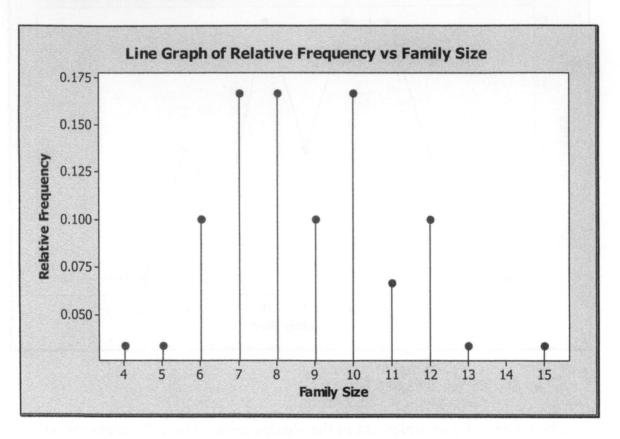

9. (a) The proportion of months with *fewer than* 2 emergency appendectomies

    $= 0.05 + 0.08 = 0.13$ (or 13%).

   (b) The proportion of months with *more than* 5 emergencies appendectomies

    $= 0.15 + 0.10 = 0.25$ (or 25%).

   (c) One way of observing whether the data set is symmetrical, is to analyze the relative frequency (or frequency) line graph for the data set. The relative frequency line graph for the data set is shown below.

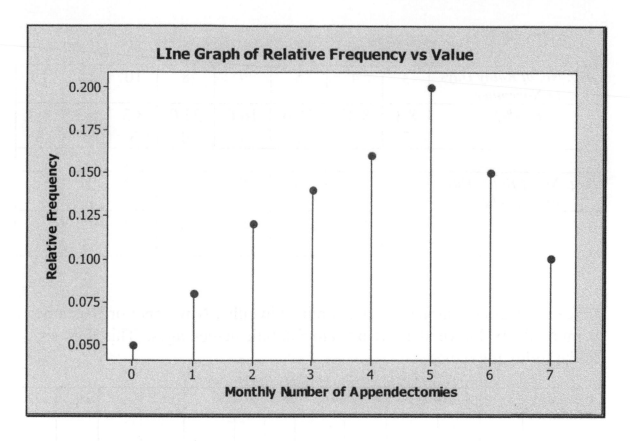

The line graph shows that the data set is skewed to the left. Thus, we cannot say that the data set is symmetric.

11. (a) Proportion of ***winning scores in the Masters*** that are ***below*** 280

$$= 0.027 + 0.027 + 0.027 + 0.027 + 0.054 + 0.108 + 0.135 + 0.081$$
$$+ 0.162 = 0.648.$$

(b) Proportion of ***winning scores in the Masters*** that are ***282 or higher***

$$= 0.027 + 0.054 + 0.027 + 0.027 + 0.027 = 0.162.$$

(c) Proportion of ***winning scores in the Masters*** that are ***between 278 and 284 inclusive***

$$= 0.081 + 0.162 + 0.108 + 0.081 + 0.027 + 0.054 + 0.027 = 0.54.$$

13. The table below shows the relative frequency table for the data values relating to the ***average number of rainy days in either November or December for the first 12 cities***. The relative frequency is expressed as

a percent (%).

Avg. No. of Rainy Days in November	3	4	5	7	8	10	11	19
f/n (%)	8.33	8.33	16.67	16.67	25.00	8.33	8.33	8.33

Avg. No. of Rainy Days in December	4	5	6	9	10	12	21	
f/n (%)	0.33	16.67	0.33	25	25	0.33	0.33	

Since the average number of days it rained in either November or December is mutually exclusive, then we need to combine percentages. This is given in the following table.

Avg. No. of Rainy Days in either Nov. or Dec.	3	4	5	6	7	8	9	10	11	12	19	21
Combined %	8.33	8.66	33.34	0.33	16.67	25	25	33.33	8.33	0.33	8.33	0.33

15. Below is the pie chart for the *various classifications of deaths on British roads in 1987*.

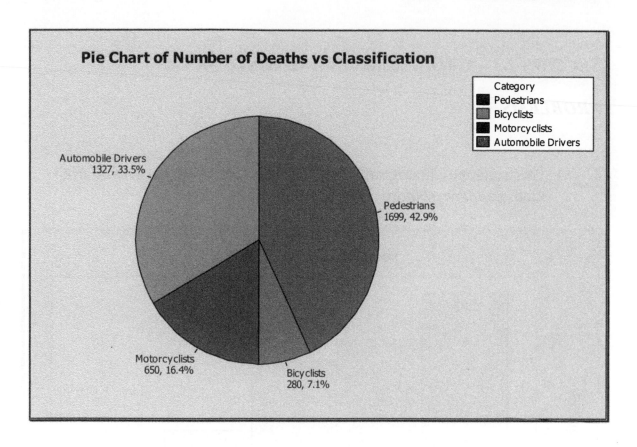

Pie Chart of Number of Deaths vs Classification

Category
Pedestrians
Bicyclists
Motorcyclists
Automobile Drivers

Automobile Drivers
1327, 33.5%

Pedestrians
1699, 42.9%

Motorcyclists
650, 16.4%

Bicyclists
280, 7.1%

SECTION 2.3 -- GROUPED DATA AND HISTOGRAMS

PROBLEMS

1. (a) The frequency histogram, with 5 classes, for the *IQ scores of the sixth graders* is shown below.

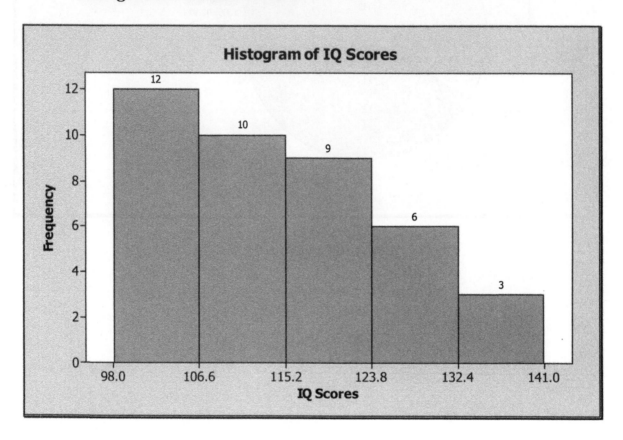

(b) The class with the largest frequency is from 98 to 106.6. The frequency for this class is $f = 12$.

(c) There is *not* a roughly equal number of data in each class interval as displayed in the graph.

(d) The histogram does not appear too symmetrical.

3. (a) Frequency histogram for **gross adjusted incomes** (in thousands of dollars) using 5 classes.

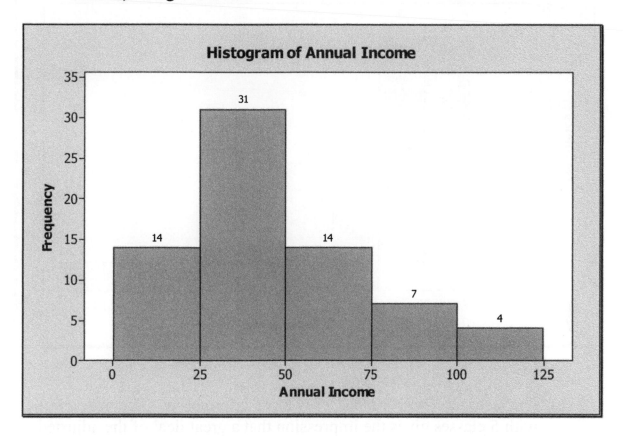

(b) Frequency histogram for **gross adjusted incomes** using 10 classes.

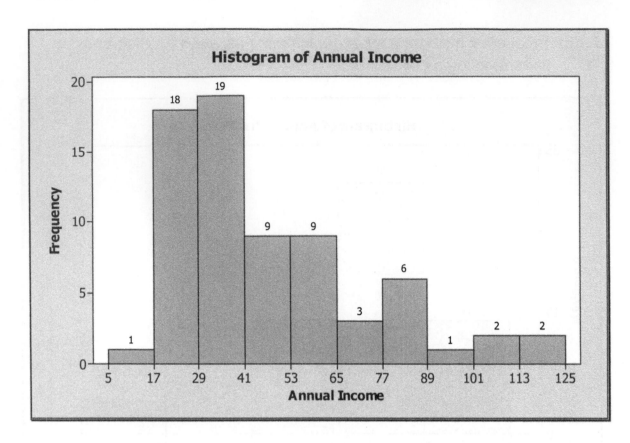

(c) The histogram with 10 classes is more informative. The histogram with 5 classes gives the impression that a great deal of the adjusted incomes is less than $50,000. However, with ten classes, we see that a majority of the adjusted incomes is below $65,000. Also, with this histogram (10 classes), one can get a better idea of the distribution for the incomes.

5. (a) The following frequency histogram for *ozone concentration* includes the class interval 3 - 5.

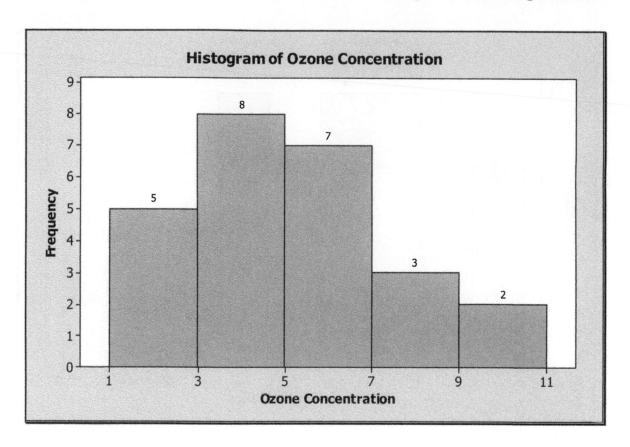

(b) The following frequency histogram for *ozone concentration* includes the class interval 2 - 3.

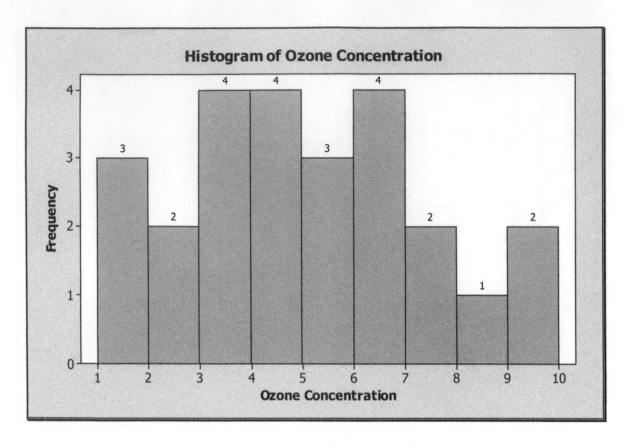

(c) The frequency histogram in (b) is more informative. For example, we can see from (b) that there were equal number of observed ozone concentrations for different interval values (classes) of size one. Observe there were an equal number of observed values in the intervals (classes) 3 - 4, 4 - 5, and 6 - 7 etc. This kind of information is not revealed in the frequency histogram in part (a).

7. (a) Class relative frequency table for *the Males and Females cholesterol levels*.

Classes	*Male (f)*	*Male (f/n)*	*Female (f)*	*Female (f/n)*
165 - 175	2	0.0377	0	0.0000
175 - 185	7	0.1321	2	0.0426
185 - 195	18	0.3400	10	0.2128
195 - 205	13	0.2453	19	0.4043
205 - 215	5	0.0943	12	0.2553
215 - 225	6	0.1132	4	0.0755
225 - 235	2	0.0377	0	0.0000
	Total, n = 53		Total, n = 47	

(b) The relative frequency polygons for the *male and female blood cholesterol levels* are shown below.

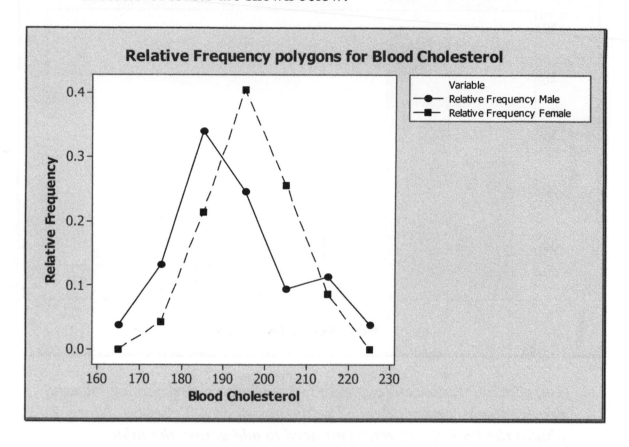

(c) From the relative frequency polygons it appears that the higher blood cholesterol levels for the males are more spread out. Also, it appears that there is a smaller chance for the blood cholesterol levels for the females to be lower than the males for levels up to 195 (approximately), a higher chance between 195 and 218 (approximately), and a lower chance between 195 and 230.

9. The relative frequency (expressed as percents) histogram for the *death rates due to* motor vehicles is shown below. This graph was constructed with 7 class intervals.

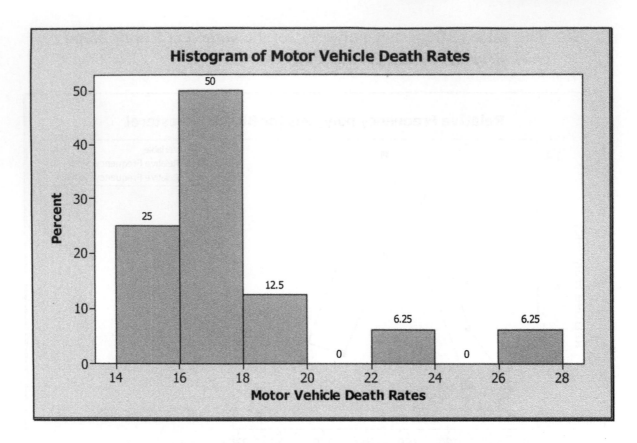

11. The relative frequency (expressed as percents) histogram for the ***total death rates*** is shown below. The histogram used 7 classes. ***Note: To obtain the total death rates you need to add across the table***.

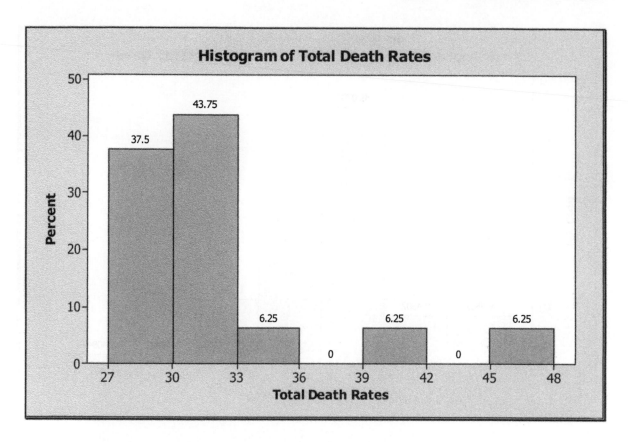

13. To construct a histogram for the ***average yearly number of rainy days***
for the listed cities, we have to use the data in the last column (annual)
of the table. A relative frequency histogram shown below used 7
classes. (Note: No specification was given as to the type of histogram-
frequency or relative frequency).

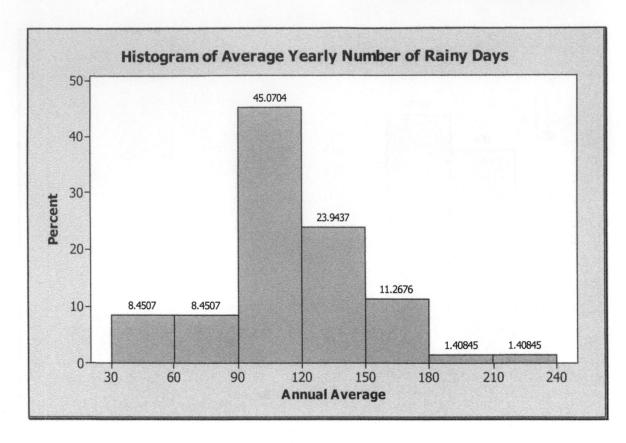

15. (a) The cumulative relative frequency (as a percentage) should add up
     to 100 since 100% of the data values for both age groups will be
     less than or equal to 240.

    (b) The completed table is shown below. The **numbers** that were filled
     in the body of the table are indicated in **bold**.

Blood Pressure less than	Ages 30 - 40 (Cumulative Percents)	Ages 50 - 60 (Cumulative Percents)
90	0.12	0.14
100	0.79	0.41
110	5.44	3.56
120	**23.55**	**11.36**
130	**53.79**	**28.05**
140	**80.36**	**48.43**
150	**92.64**	**71.27**
160	**97.36**	**81.26**
170	**99.13**	**89.74**
180	**99.84**	**94.53**
190	**99.96**	**97.27**

200	**100**	**98.5**
210	**100**	**98.91**
220	**100**	**99.59**
230	**100**	**99.86**
240	100	100

(c) The age group from 30 to 40 years of age appears to have smaller values. Observe for this age group, 100% of the data values were less than 200. For the age group from 50 to 60 years of age, 100% of the data values were less than 240.

(d) The figure below shows the **Ogives** (cumulative relative frequency graph)for the blood pressure for the two groups of men.

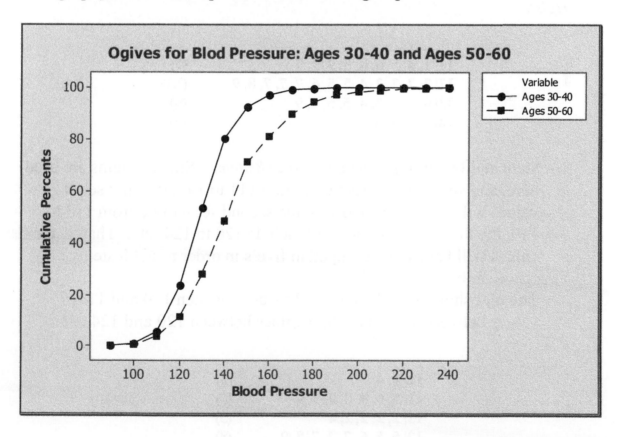

SECTION 2.4 -- STEM AND LEAF PLOTS

PROBLEMS

1. Observe that there are 36 data values ranging from 111 to 147.

(a) Stem and leaf plot for the data using *4 stems*. Since 4 stems were used, and the data set ranged from 111 to 147, then the stems will be in the 110s, 120s, 130s, and 140s. The plot says that there are 9 values with a stem of 11, 13 values with a stem of 12, 9 values with a stem of 13, and 5 values with a stem of 14. That is, the last column gives the number of values in each stem. *Note: your text has these numbers to the far right*.

11| 1, 4, 5, 6, 8, 8, 9, 9, 9   (9)
12| 2, 2, 2, 2, 4, 5, 5, 6, 7, 7, 7, 8, 9   (13)
13| 0, 2, 2, 3, 4, 5, 5, 7, 9   (9)
14| 1, 1, 4, 6, 7   (5)

(b) Stem and leaf plot for the data using *8 stems*. Since 8 stems are used here, and since the values range from 111 to 147, the first set of values will be from 110 to 114, the second set will be from 115 to 119, the third set of values will include 120 to 124, etc. That is, the values will have to be grouped in five's in order to get 8 stems.

The plot shows that there were 2 values between 110 and 114, 7 values between 115 and 119, 5 values between 120 and 124, etc.

11| 1, 4   (2)
11| 5, 6, 8, 8, 9, 9, 9   (7)
12| 2, 2, 2, 2, 4   (5)
12| 5, 5, 6, 7, 7, 7, 8, 9   (8)
13| 0, 2, 2, 3, 4   (5)
13| 5, 5, 7, 9   (4)
14| 1, 1, 4   (3)
14| 6, 7   (2)

3. A stem and leaf plot for the *ages of the 43 patients admitted to the emergency ward* of a certain adult hospital is shown below. The five year interval of ages that contains the largest number of data points is from 15 to 19 years.

1| 4	(1)
1| 5, 6, 6, 7, 7, 7, 7, 8, 8, 8, 9, 9, 9, 9	(14)
2| 0, 0, 0, 0, 1, 2, 2, 2, 3, 4	(10)
2| 5, 7, 7, 9	(4)
3| 0, 1, 1, 2, 3	(5)
3|	(0)
4| 0, 4, 4	(3)
4| 5	(1)
5| 1, 3	(2)
5| 5	(1)
6| 1	(1)
6|	(0)
7|	(0)
7| 9	(1)

5. (a) A stem and leaf plot for *New York City's daily revenue from parking meters during 30 days in 2002* is shown below.

3|2	(1)
4|	(0)
5|2, 7, 8, 9	(4)
6|5, 8, 8	(3)
7|1, 4, 5, 5, 7, 8, 9	(7)
8|0, 1, 3, 3, 3, 4, 8, 8	(8)|
9|0, 3, 4, 7	(4)
10|0, 4, 8	(3)

(b) The value of $32,000 seems "suspicious". It is small relative to the rest of the values.

7. (a) A stem and leaf plot for the *winning scores for the first 25 super bowl games in professional football* is shown below.

```
1| 4                          (1)
1| 6, 6, 6                    (3)
2| 0, 0, 1, 3, 4, 4          (6)
2| 6, 7, 7, 7                (4)
3| 1, 2, 3                   (3)
3| 5, 5, 8, 8, 9            (5)
4| 2                          (1)
4| 6                          (1)
5|                            (0)
5| 5                          (1)
```

(b) A stem and leaf plot for the *losing scores for the first 25 super bowl games in professional football* is shown below.

```
0| 3                                (1)
0| 6, 7, 7, 7, 7, 9                 (6)
1| 0, 0, 0, 0, 0, 0, 3, 4, 4       (9)
1| 6, 6, 7, 7, 9, 9                 (6)
2| 0, 1                            (2)
2|                                  (0)
3| 1                                (1)
```

(c) A stem and leaf plot for the *amounts by which the winning teams outscored the losing teams for the first 25 super bowl games in professional football* is shown below.

```
0| 1, 3, 4, 4, 4                   (5)
0| 5, 7, 9                         (3)
1| 0, 0, 2                         (3)
1| 6, 7, 7, 7, 8, 9, 9            (7)
2| 1, 2                           (2)
2| 5, 9                           (2)
3| 2                               (1)
3| 6                               (1)
4|                                 (0)
4| 5                               (1)
```

9. (a) There are 6 data values in the forties (42, 42, 45, 48, 48, 48).

   (b) The number of data values greater than 50 = 14.

   The total number of values, n = 32.

   Thus, the percentage of data values larger than 50 = (14/32)×100%

$$= 43.75\%.$$

   (c) The number of data values having their ones (unit) digit equal to 1 = 4.

   The total number of values, n = 32.

   Thus, the percentage of data values having their ones digit equal to 1

$$= (4/32)\times100\% = 12.5\%.$$

11. (a) School B had the "high scorer". A single student (from school B) scored 100 (points).

   (b) School A had the "low scorer". A single student (from school A) scored the lowest of 50 (points).

   (c) School A did better on the examination since 21 students from school A scored 70 and above as compared to only 16 students from school B.

   (d) A stem and leaf plot for all 48 values is shown below.

```
 5| 0, 3, 5, 7                              (4)
 6| 2, 5, 5, 8, 8, 9, 9                      (7)
 7| 0, 2, 3, 4, 6, 7, 7, 8, 8, 9, 9         (11)
 8| 0, 2, 3, 3, 5, 5, 6, 6, 6, 7, 7, 8, 8, 9  (14)
 9| 0, 0, 1, 3, 5, 5, 5, 6, 6, 8, 8         (11)
10| 0                                       (1)
```

SECTION 2.5 -- SETS OF PAIRED DATA

PROBLEMS

1. (a) The scatter diagram for the ***number of defective parts*** versus ***daily midday temperatures*** is shown below.

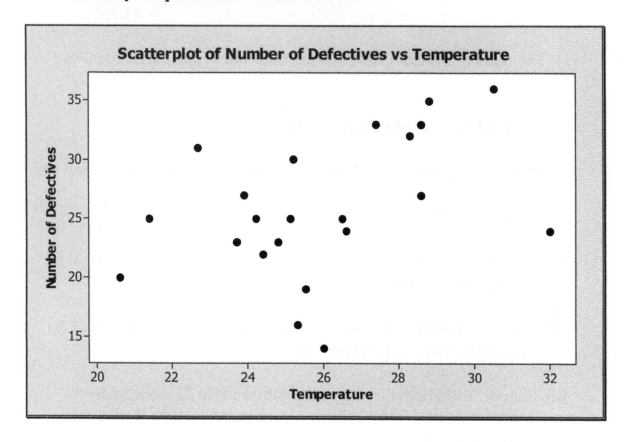

(b) The scatter diagram seem to have some predictive uses since, generally, it seems that larger values of the ***midday temperature*** are associated with larger values for the ***number of defective parts*** produced.

(c) Based on a "fit by eye" line, the best guess for the number of defective parts produced when the temperature reading is 24 degrees Celsius would be close to 24. You can obtain this approximate value by drawing a line vertically from the temperature axis at the point (approximate) 24 degrees to the "fit by eye" line. From this point on the line, draw a horizontal line to the vertical axis. This horizontal line will intersect the vertical axis at approximately a value of 24. This is shown below.

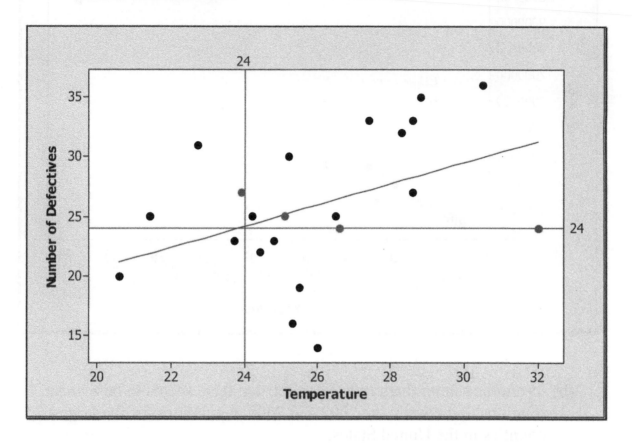

3. (a) The scatter diagram for the 2000 and 2002 populations of some of the largest counties in the United States is shown below.

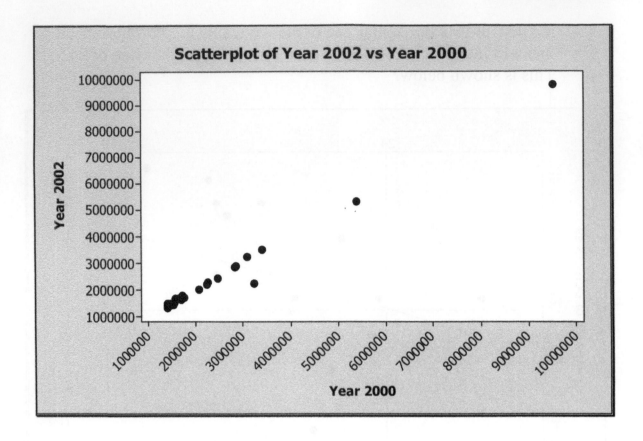

(b) From the scatter diagram we can see that there seems to be a linear relationship between the 2002 and 2000 populations for the largest 25 counties in the United States.

5. (a) The scatter diagram for the ***attention span in minutes*** versus the ***IQ score*** is shown below.

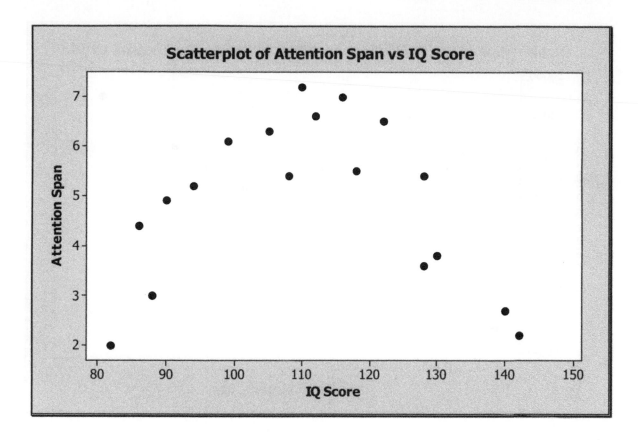

(b) From the scatter diagram we see that low IQ scores are associated with a low attention span. As the IQ score increases, the attention span increases to a maximum between an IQ score of 110 and 120. Beyond this range, as the IQ score increases, the attention span decreases. Thus, one can infer that humans with low and high IQ scores will have a low attention span. Humans with IQ scores between 110 and 120 (approximately) will have a maximum attention span of approximately 7 minutes. Humans with high IQ scores will have low attention span.

7. (a) The scatter diagram for the *1996 per capita income* versus the *1994 per capita income* for 12 different U.S. metropolitan areas is shown below.

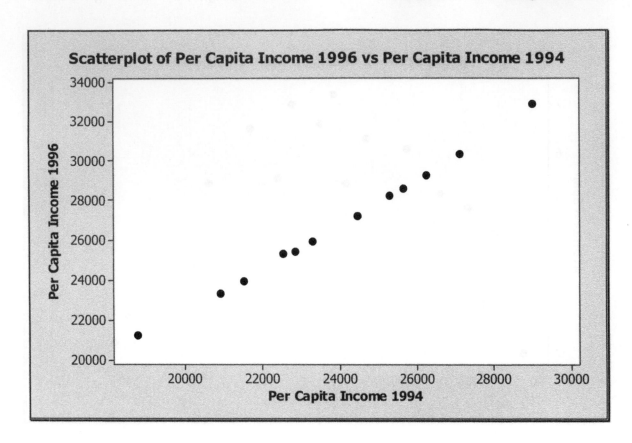

Scatterplot of Per Capita Income 1996 vs Per Capita Income 1994

(b) From the scatter diagram in part (a), when the 1994 per capita income of the residents of San Diego was $22,111, the annual 1996 per capita income was approximately $24,700.

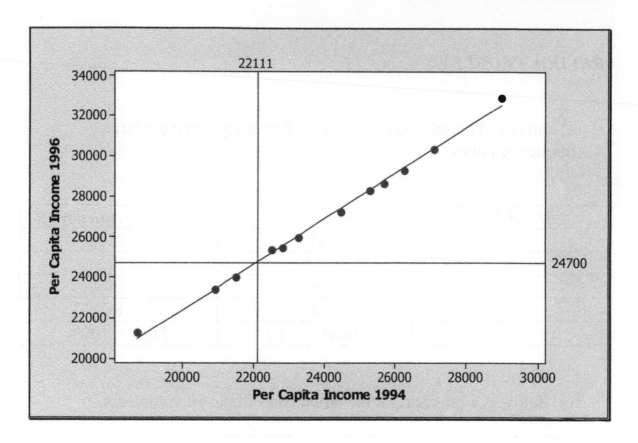

REVIEW PROBLEMS

1. (a) and (b). This table shows both the frequency and the relative frequency tables.

Blood Type	Frequency	Relative Frequency (%)
O	19	38
A	19	38
B	8	16
AB	4	8
	Total, n = 50	*Total* = 100%

(c) Below is a pie chart for the *blood types* of the 50 volunteers.

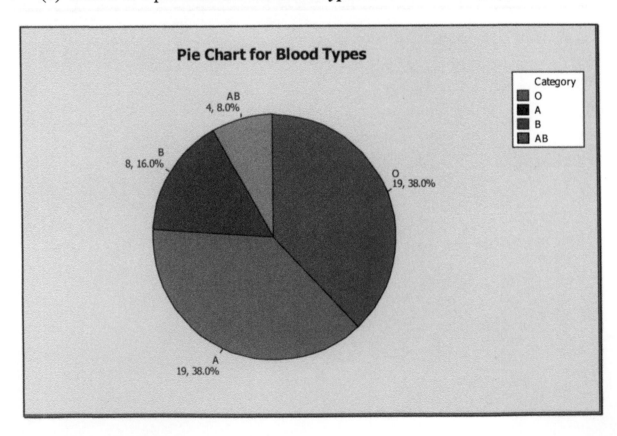

3. (a) Total number of suicides that were reported over the 14 years

$$= 9 + 19 + 17 + 20 + 15 + 11 + 8 + 2 + 3 + 5 + 3 = 112.$$

(b) A histogram for the ***number of female suicides in 8 German states over a 14 year period*** is shown below.

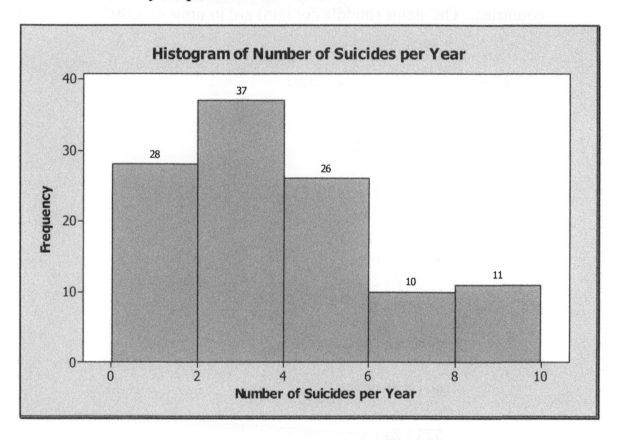

5. (a) A sample frequency table for 10 data values that is symmetric and has 5 distinct values is shown below.

Value	-2	-1	0	1	2
Frequency	1	2	4	2	1

(b) A sample frequency table for 10 data values that is symmetric and has 4 distinct values is shown below.

Value	1	2	3	4
Frequency	2	3	3	2

(c) The data set in (a) is symmetric about 0 and the data set in (b) is symmetric about 2.5 {(2+3)/2}, since we have an even set of numbers and so we take the average of the middle two).

7. (a) A back-to-back stem and leaf diagram for the *2000* and *2002* amounts (millions of $) invested in the United States by selected European countries. The stems (middle column) are in units of 1,000.

<table>
<tr><td align="center">**2000**</td><td></td><td align="center">**2002**</td></tr>
</table>

```
319 | 0 | 259
    | 1 | 924
665 | 2 |
007 | 3 | 416, 439
025 | 4 | 739
068 | 5 |
576 | 6 | 695
    | 7 | 212
875 | 8 |
          .
787 | 14 |
          .
991 | 21 | 989
          .
523 | 25 |
    | 26 | 179
          .
    | 34 | 349
          .
930 | 58 |
          .
719 | 64 |
          .

          .

          .
    | 283 | 317
```

(b) The scatter diagram for the *2000* and *2002* amounts (millions of $) invested in the United States by selected European countries is shown below.

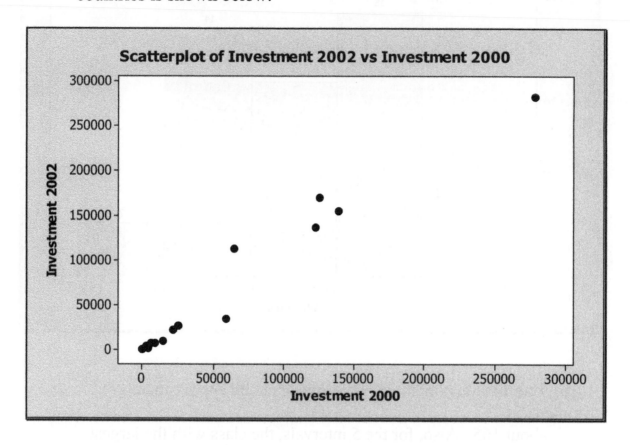

9. (a) The histogram of the *weights of the last 50 entries in Appendix A* is shown below. This histogram was constructed using 5 classes.

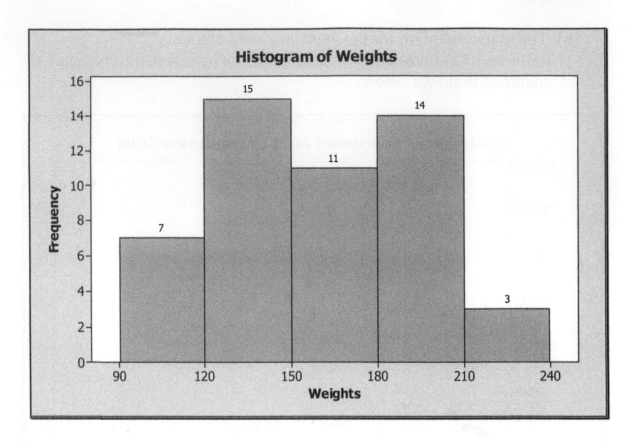

(b) From the histogram in (a), it appears to be approximately symmetrical
    about 165. Also, for the 5 intervals, the class with the largest
    frequency is between 120 and 150.

11. The scatter diagram for the ***weight versus the blood pressure for
    the last 50 entries in Appendix A*** is shown below.

    From the scatter diagram, we can observe that generally
    that larger values of weights are associated with larger values of
    blood pressure and smaller values weights are associated with
    smaller values of blood pressure. This will suggest that there is a
    positive association between these two variables.

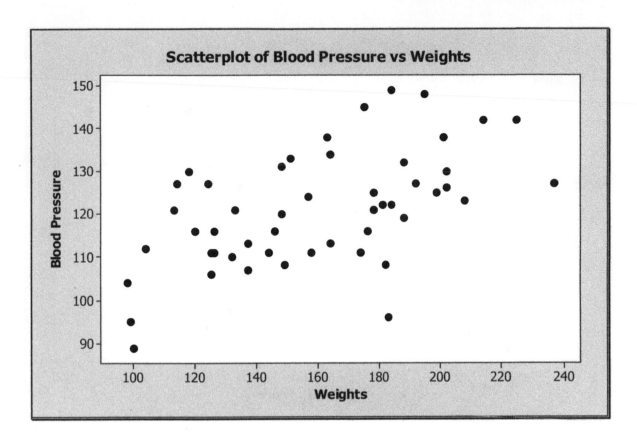

13. The scatter diagram for the **verbal** versus the **math SAT scores for the14 high school seniors** is shown below.

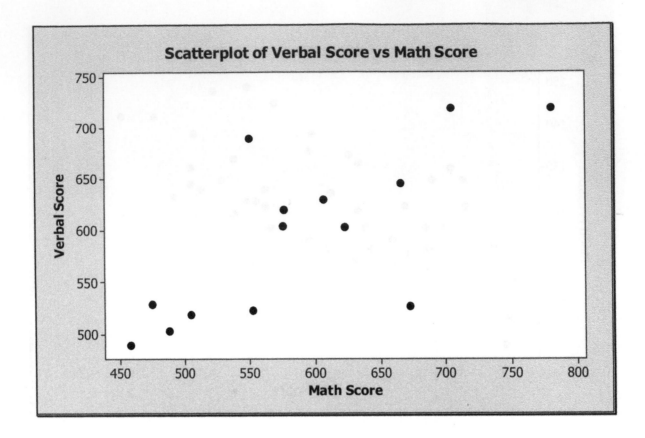

From the scatter diagram, high verbal scores generally tend to go with high

math scores. This would indicate that there is a positive association between the two variables.

15. (a) A stem and leaf plot for the ***normal precipitation amounts for April*** is shown below.

>   0 | .27, .78, .93
>   1 | .19, .31, .49, .53, .81
>   2 | .30, .92, .93
>   3 | .07, .21, .32, .39, .66, .68, .81
>   4 | .02, .11, .43, .50
>   5 | .35, .41

   (b) A histogram for the ***annual amounts of precipitation*** is shown below.

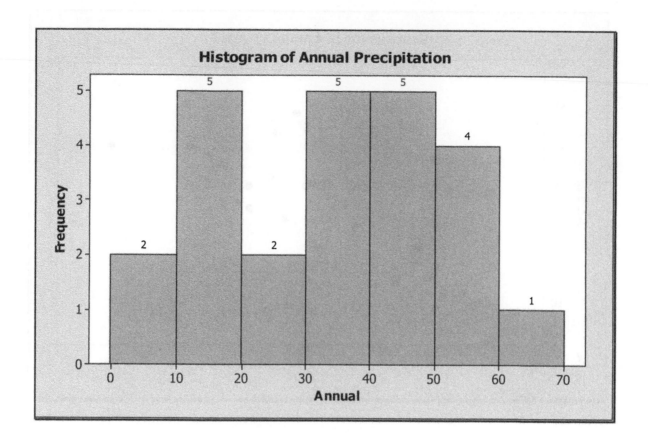

(c) A scatter diagram for the ***annual precipitation*** versus the ***normal precipitation amounts for April*** is shown below.

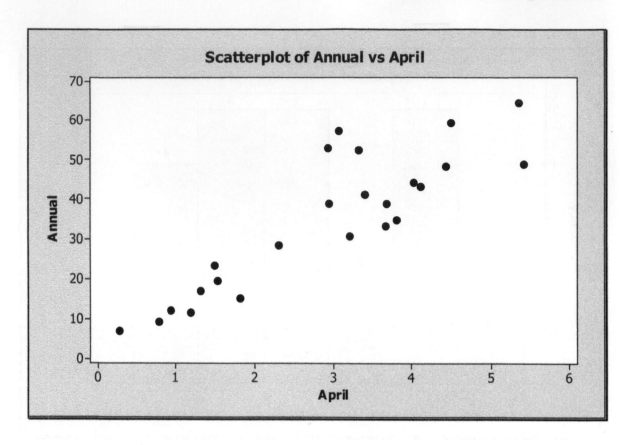

17. (a) Seems to be a general increase for the Japanese imports and a general decrease then an increase for the German imports.

    (b) The scatter diagram relating the Japanese and German car imports since 1970 is shown below.

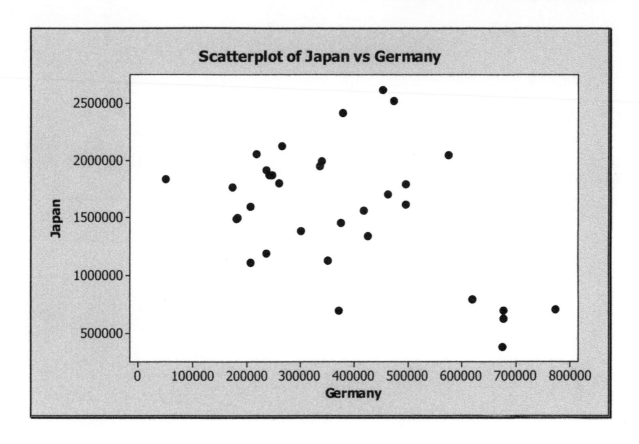

Chapter 3   USING STATISTICS TO SUMMARIZE DATA

SECTION 3.2 -- THE SAMPLE MEAN, DEVIATIONS

PROBLEMS

1. The sample mean, $\bar{x} = (87 + 63 + 91 + ... + 75 + 88)/15 = 79.73$.

3. (a) The sample mean of the average number of inches of precipitation

    $$\bar{x} = (35.74 + 31.50 + 11.43 + ... + 29.13 + 28.61)/13 = 33 \text{ (inches)}.$$

    (b) The sample mean of the average number of days of precipitation

    $$\bar{x} = (134 + 83 + 95 + ... + 81 + 85)/13 = 102.38 \text{ (days)}.$$

5. The sample mean, $\bar{x} = (164 + 165 + 157 + ... + 145 + 132)/12$
    $= 150.58$ (law enforcement officers killed).

7. The sample mean $\bar{x} = (148 + 102 + 95 + 88 + 74 + 83 + 48 + 62)/8$
    $= 97.5$ (reported cases of tetanus).

9. (a) The sample mean $\bar{x} = (1 + 2 + 4 + 7 + 10 + 12)/6 = 6$.

    (b) The sample mean $\bar{x} = (3 + 6 + 12 + 21 + 30 + 36)/6 = 18$.

    (c) The sample mean $\bar{x} = (6 + 7 + 9 + 12 + 15 + 17)/6 = 11$.

11. The sample mean $\bar{x} = (6 + 13 + 5 + 7 + ... + 6 + 9 + 8)/12 = 6.5$
    (number of fires)

13. The sample mean $\bar{x} = 1/2 \times 10 + 1/2 \times 20 = 15$.

15. The sample mean $\bar{x} = (1/2 \times 10 + 1/6 \times 20 + 1/3 \times 30) = 18.3333$.

17. The combine sample size $n = 20 + 30 = 50$. The weights are $\dfrac{20}{50}$ and $\dfrac{30}{50}$. Thus, the sample mean can be computed from

$\bar{x} = \dfrac{20}{50}(42400) + \dfrac{30}{50}(33600) = 37120$. That is, the mean salary for the 50 engineers is \$37,120.

19. Recall, the respective means were $\bar{x} = 6$; $\bar{x} = 18$; $\bar{x} = 11$.

Data Set 1	Deviations	Data Set 2	Deviations	Data Set 3	Deviations
1	-5	3	-15	6	-5
2	-4	6	-12	7	-4
4	-2	12	-6	9	-2
7	1	21	3	12	1
10	4	30	12	15	4
12	6	36	18	17	6
Total	0	Total	0	Total	0

SECTION 3.3 -- THE SAMPLE MEDIAN

PROBLEMS

1. (a) First arrange the data set in ascending order. Since the data set is even (n = 12), the median will be the average of the two middle numbers (average of the 6th and 7th in the ordered set). That is, the median m = (6540 + 6620)/2 = 6580 yards.

   (b) The sample mean, $\bar{x}$ = 6545 yards.

3. The median for the data set $2x_i + 3 = 2(10) + 3 = 23$.

5. First arrange the data sets in ascending order. Since the data set is odd (n = 13) for both the male and female suicide rates, the median will be the middle number (7th in the ordered sets) in both cases.

   (a) The sample median of the male suicide rates, m = 22.

   (b) The sample median of the female suicide rates, m = 8.1.

   (c) The sample mean of the male suicide rates, $\bar{x}$ = 23.68.

   (d) The sample mean of the female suicide rates, $\bar{x}$ = 9.68.

7. First arrange the data set in ascending order. Since the data set is odd (n = 13), the median will be the middle value (7th in the ordered set). That is, the median of the average annual number of inches of precipitation for the cities, m = 31.5.

9. (a) First arrange the data set in ascending order. Since the data set is even (n = 16), the median will be the average of the two middle values (average of the 8th and 9th values in the ordered set). That is, the median of the death rates due to falls, m = (5.1 + 5.2)/2 = 5.15.

   (b) First arrange the data set in ascending order. Since the data set is even (n = 16), the median will be the average of the two middle values (average of the 8th and 9th values in the ordered set). That is, the median of the death rates due to poisoning, m = (3.4 + 3.5)/2 = 3.45.

(c) First arrange the data set in ascending order. Since the data set is even (n = 16), the median will be the average of the two middle values (average of the 8th and 9th values in the ordered set). That is, the median of the death rates due to drowning, m = (1.5 + 1.5)/2 = 1.5.

11. To compute the values for part (a) and (b) we need to use the largest possible values in the intervals shown on the histogram. The largest values in the intervals are the right end points. The table below shows these values (x) with their corresponding frequencies (f).

x	16	17	18	19	20	21	22	23	24	25
f	1	3	3	4	6	4	3	5	2	3

(a) Observe that the sum of the frequencies is 34. Thus, the sample mean $\bar{x}$ = (16×1 + 17×3 + 18×3 + ... + 24×2 + 25×3)/34 = 20.74

(b) Since the data set is even (n = 34), the sample median will be the average of the two middle numbers (17th and 18th). Thus the sample median, m = (20 + 21)/2 = 20.5.

To compute the values for part (c) and (d) we need to use the smallest possible values in the intervals shown on the histogram. The smallest values in the intervals are the left end points. The table below shows these values (x) with their corresponding frequencies (f).

x	15	16	17	18	19	20	21	22	23	24
f	1	3	3	4	6	4	3	5	2	3

(c) Observe that the sum of the frequencies is 34. Thus, the sample mean $\bar{x}$ = (15×1 + 16×3 + 17×3 + ... + 23×2 + 24×3)/34 = 19.74.

(d) Since the data set is even (n = 34), the sample median will be the average of the two middle numbers (17th and 18th). Thus the sample median, m = (19 + 20)/2 = 19.5.

(e) The actual sample mean $\bar{x} = 20.22$ and the actual sample median $m = 20.05$.

13. From the frequency distribution, the median class for both groups (driver
    with helmet and driver without helmet) is 0. Since the middle values for
    these odd sets of data will be in the median class, then the median for
    both groups is 0.

15. The sample size n = 14.

    (a) The sample mean $\bar{x} = 32.52$.

    (b) The sample median, m = 24.25.

       You would expect the answers to be inconsistent since we do not
       know the actual number of females in a given occupation and the
       total for that occupation. For the mean to be consistent with 44.4%
       we need to know the weights (number of females in an
       occupation/total in the given occupation) for the different
       occupations such that we could use these weights to compute the
       weighted mean.

17. Arrange the data sets in ascending order and since the sample size
    n = 11, the sample medians will be the 6th values in the ordered sets.

    (a) The sample median of the men's median age m = 26.8.

    (b) The sample median of the women's median age m = 25.

SECTION 3.3.1 -- SAMPLE PERCENTILES

PROBLEMS

1. (a) First arrange the data set in ascending order. Since $n = 75$ and $p = 0.8 \Rightarrow np = 75 \times 0.8 = 60$ which is an integer. Thus the sample 80-percentile will be the average of the 60th and 61st data values in the ordered set.

   (b) First arrange the data set in ascending order. Since $n = 75$ and $p = 0.6 \Rightarrow np = 75 \times 0.6 = 45$ which is an integer. Thus the sample 60-percentile will be the average of the 45th and 46th data values in the ordered set.

   (c) First arrange the data set in ascending order. Since $n = 75$ and $p = 0.3 \Rightarrow np = 75 \times 0.3 = 22.5$ which is not an integer. Round up to the next integer (23). Thus the sample 30-percentile will be the 23rd data value in the ordered set.

3. Since the data set is 1, 2, 3, ..., n we can assume that the values are integers and thus have already been ordered.

   (a) $n = 100$, $p = 0.95 \Rightarrow np = 100 \times 0.95 = 95$. Since np is an integer then the sample 95-percentile for the data set will be the average of the 95th and 96th values. Hence, the sample 95-percentile is $(95 + 96)/2 = 95.5$.

   (b) $n = 101$, $p = 0.95 \Rightarrow np = 101 \times 0.95 = 95.95$. Since np is not an integer round up to the next integer (96). Thus the sample 95-percentile for the data set will be located in the 96th position. Hence, the sample 95-percentile is 96.

5. First arrange the data set in ascending order.

   (a) $n = 12$, $p = 0.9 \Rightarrow np = 12 \times 0.9 = 10.8$. Since np is not an integer round up to the next integer (11). Thus the sample 90-percentile for the dentist's rates per 100,000 will be located in the 11th position. Hence, the sample 90-percentile is 70.

(b) $n = 12$, $p = 0.5 \Rightarrow np = 12 \times 0.5 = 6$. Since np is an integer (6) then the sample 50-percentile will be the average of the 6th and 7th values in the ordered set. Thus the sample 50-percentile for the dentist's rates per 100,000 will be $(56 + 60)/2 = 58$.

(c) $n = 12$, $p = 0.1 \Rightarrow np = 12 \times 0.1 = 1.2$. Since np is not an integer round up to the next integer (2). Thus the sample 10-percentile for the dentist's rates per 100,000 will be located in the 2nd position. Hence, the sample 10-percentile is 52.

7. Given that the sample 100p-percentile of a data set is 230. Multiplying each value by a positive constant c will not change the 100p-percentile of 230 to 230c. Thus the new sample 100p-percentile will still be 230c.

9. First arrange the first 51 rates in ascending order.

$n = 51$, $p = 0.25 \Rightarrow np = 51 \times 0.25 = 12.75$. Since np is not an integer round up to the next integer (13). Thus the sample 25-percentile for the fatality rates per 100 million will be located in the 13th position. Hence, the sample 25-percentile or the first quartile is 1.27.

$n = 51$, $p = 0.5 \Rightarrow np = 51 \times 0.5 = 25.5$. Since np is not an integer round up to the next integer (26). Thus the sample 50-percentile for the fatality rates per 100 million will be located in the 26th position. Hence, the sample 50-percentile or the second quartile (median) is 1.55.

$n = 51$, $p = 0.75 \Rightarrow np = 51 \times 0.75 = 38.25$. Since np is not an integer round up to the next integer (39). Thus the sample 75-percentile for the fatality rates per 100 million will be located in the 39th position. Hence, the sample 75-percentile or the third quartile is 1.84.

11. Since the distribution is symmetric, then the distance from the median to the third quartile will equal the distance from the first quartile to the median.

Hence, the distance will be 55 – 40 = 15. Hence the value of the first quartile will be 40 – 15 = 25.

SECTION 3.4 -- THE SAMPLE MODE

PROBLEMS

1. (a) Sample mode is 9 -- data set B

   (b) Sample mean is 9 -- data set C

   (c) Sample median is 9 -- data set A

3. (a) Weight -- the mode is 126.

   (b) Blood pressure -- the modes are 102, 110, and 114.

   (c) Cholesterol -- the mode is 196.

5. A data set for which the sample mean is 10, the sample median is 8, and the sample mode is 6 is 6, 6, 8, 9, 21. Another example of such a data set is 6, 6, 8, 14, 15, 15.

7. Let **x** be the number of loops and **f** the frequency for the number of loops. The table below shows the frequency distribution for this data.

x	2	4	6	8	12
f	1	4	5	6	1

   (a) Observe from the table that 8 loops occur the most with a frequency of 6. Thus the sample mode of the number of loops ran by these joggers is 8.

   (b) Since the track is a quarter of a mile around, then a jogger must complete 4 loops to cover one mile. Thus the sample mode of the distances ran by these joggers is 8/4 miles or 2 miles. We have to divide 8 by 4 since the mode of the number of loops ran by the joggers

      is 8.

SECTION 3.5 -- SAMPLE VARIANCE AND SAMPLE STANDARD DEVIATION

PROBLEMS

1. The sample mean $\bar{x} = (26.3 + 26.2 + 26.4 + 26.3 + 25.9)/5 = 26.22$.

    The sample variance s^2 can be computed from the table below.

x_i	$(x - \bar{x})$	$(x - \bar{x})^2$
26.3	0.08	0.0064
26.2	-0.02	0.0004
26.4	0.18	0.0324
26.3	0.08	0.0064
25.9	-0.32	0.1024
		$\sum_{i=1}^{5} (x_i - \bar{x})^2 = 0.148$

Thus the sample variance $s^2 = (0.148)/(5 - 1) = 0.037$.

Alternatively, you can find the sample variance by computing $\bar{x} = 26.22$, $\sum_{i=1}^{5} x_i^2 = 3{,}437.59$, $n\bar{x}^2 = 3{,}437.44$, $n = 5$, so the sample variance $s^2 = (3{,}437.59 - 3{,}437.44)/(5 - 1) = 0.0375$.

3. (a) U.S. Open: $\bar{x} = 278.2$, $\sum_{i=1}^{10} x_i^2 = 774{,}008$, $n\bar{x}^2 = 773{,}952.4$, $n = 10$, so the sample variance $s^2 = (774{,}008 - 773{,}952.4)/(10 - 1) = 6.18$.

    (b) Masters: $\bar{x} = 280.9$, $\sum_{i=1}^{10} x_i^2 = 789{,}109$, $n\bar{x}^2 = 789{,}048.1$, $n = 10$, so the sample variance $s^2 = (789{,}109 - 789{,}048.1)/(10 - 1) = 6.77$.

5. Dentists: $\bar{x} = 75463$, $\sum_{i=1}^{9} x_i^2 = 51{,}969{,}808{,}473$, $n\bar{x}^2 = 51{,}251{,}979{,}321$, $n = 9$, so the sample variance

$$s^2 = (51{,}969{,}808{,}473 - 51{,}251{,}979{,}321)/(9 - 1) = 89{,}728{,}644.$$

7. Using technology (MINITAB), $s^2 = (69{,}426{,}236)^2$. Note: Squaring such a large number yields a number with 16 digits).

9. Using technology (MINITAB), $s^2 = (5{,}642{,}159)^2$. Note: Squaring such a large number yields a number with 16 digits).

11. (a) $\bar{x} = 3$, $\sum\limits_{i=1}^{5} x_i^2 = 55$, $n\bar{x}^2 = 45$, $n = 5$, so

    the sample variance $s^2 = (55 - 45)/(5 - 1) = 2.5$ and the sample standard deviation $s = 1.58$.

    (b) $\bar{x} = 8$, $\sum\limits_{i=1}^{5} x_i^2 = 330$, $n\bar{x}^2 = 320$, $n = 5$, so

    the sample variance $s^2 = (330 - 320)/(5 - 1) = 2.5$ and the sample standard deviation $s = 1.58$.

    (c) $\bar{x} = 13$, $\sum\limits_{i=1}^{5} x_i^2 = 855$, $n\bar{x}^2 = 845$, $n = 5$, so

    the sample variance $s^2 = (855 - 845)/(5 - 1) = 2.5$ and the sample standard deviation $s = 1.58$.

    (d) $\bar{x} = 6$, $\sum\limits_{i=1}^{5} x_i^2 = 220$, $n\bar{x}^2 = 180$, $n = 5$, so

    the sample variance $s^2 = (220 - 180)/(5 - 1) = 10$ and the sample standard deviation $s = 3.16$.

    (e) $\bar{x} = 30$, $\sum\limits_{i=1}^{5} x_i^2 = 5{,}500$, $n\bar{x}^2 = 4{,}500$, $n = 5$, so

    the sample variance $s^2 = (5{,}500 - 4{,}500)/(5 - 1) = 250$ and the sample standard deviation $s = 15.81$.

13. (a) First 50 values: $\bar{x} = 115.8$, $\sum\limits_{i=1}^{5} x_i^2 = 678{,}922$, $n\bar{x}^2 = 670{,}482$,

    $n = 50$, so the sample variance $s^2 = (678{,}922 - 670{,}482)/(50 - 1)$

=172.13.

(b) Last 50 values: $\bar{x} = 120.98$, $\sum\limits_{i=1}^{5} x_i^2 = 740{,}577$, $n\bar{x}^2 = 731{,}808$,

   $n = 50$, so the sample variance $s^2 = (740{,}577 - 731{,}808)/(50 - 1)$
   $= 178.96$.

The values of the statistics for the two data sets are similar. This is not surprising since we can consider these as two random samples each of size 50. Also, if the 200 values in the data set is a truly random sample, then one would not expect much variability within subgroups from the data set.

15. $\bar{x} = 1056{,}300$, $\sum\limits_{i=1}^{5} x_i^2 = 9{,}193{,}700$, $n\bar{x}^2 = 8{,}925{,}312$,

   $n = 8$, so the sample variance $s^2 = (9{,}193{,}700 - 8{,}925{,}312)/(8 - 1)$
   $= 38341.1429 \Rightarrow$ the sample standard deviation $s = 195.8089$ in thousands. That is, $s = 195{,}808.9$.

17. (a) $\bar{x} = 3.612$, $\sum\limits_{i=1}^{5} x_i^2 = 547.115$, $n\bar{x}^2 = 521.862$, $n = 40$, so the sample

   variance $s^2 = (547.115 - 521.862)/(40 - 1) = 0.6475 \Rightarrow$ the sample standard deviation $s = 0.8047$.

   (b) The range = maximum value - minimum value = 5.12 - 2.35 = 2.77.

   (c) The interquartile range = 75-percentile - 25-percentile

$$= 4.213 - 2.973 = 1.24.$$

SECTION 3.6 -- NORMAL DATA SETS AND THE EMPIRICAL RULE

PROBLEMS

1. (a) The histogram using 5 classes is shown below.

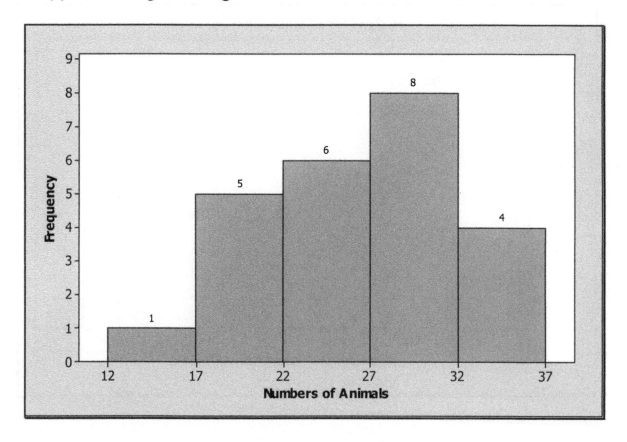

(b) The sample mean $\bar{x} = 25.75$.

(c) The sample median m = 26.5. *Note:* First you need to arrange the data set in ascending order in order to determine the median.

(d) Yes. Since the sample mean and the sample median are approximately equal, and from the shape of the histogram in part (a), we can say that the data is approximately normal.

3. (a) The sample mean $\bar{x} = 7.094$.

(b) The sample median m = 7.

5. (a) The sample mean $\bar{x}$ = $168,045.

(b) The sample median m = $172,500.

(c) The histogram for the home prices is shown below.

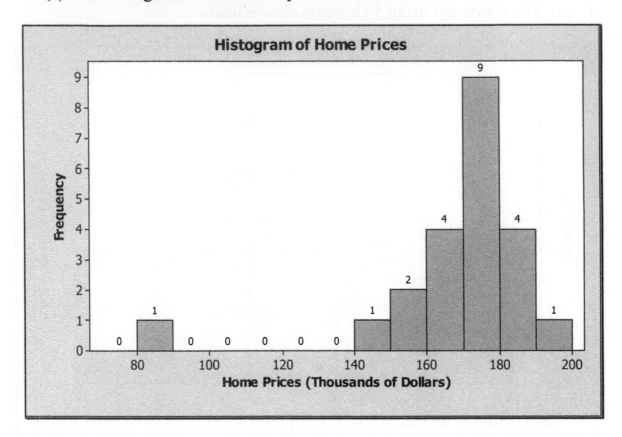

(d) No. *Note:* This response is based on using 13 classes as shown in the histogram.

NOTE: The data is approximately normal if we ignore the one outlying point. This is shown by the histogram below using 16 classes.

7. Given the sample mean $\bar{x}$ = 125.70 and the sample standard deviation s = 15.58 for the weights of the female health club members. Since the stem and leaf plot suggests that the data values are approximately normally distributed, then approximately 95% of the values will lie between 94.54 and 156.86 since this interval represents two standard deviations from the mean.

Since 92 values lie in the interval, then the actual proportion of values that are in the interval is $(92/97)\times100\% = 94.85\%$.

9. If the histogram is skewed to the right then the sample mean will be larger than the sample median.

11. (a) A histogram using seven classes is shown below for the personal income per capita for the first the first 25 states.

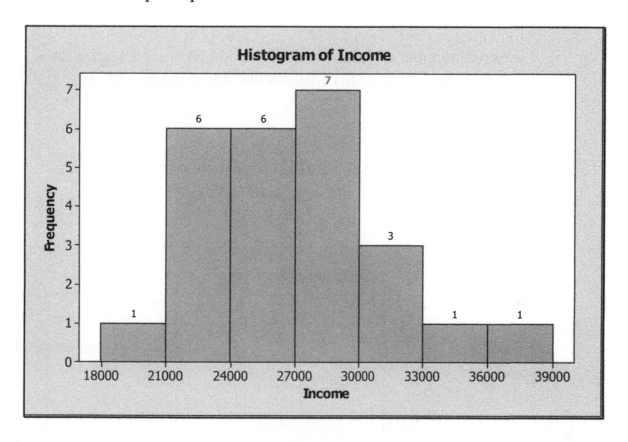

(b) The sample mean $\overline{x} = \$27,127$.

(c) The sample median $m = \$26,646$.

(d) The sample variance is $s^2 = 19,051,196$ (square $).

(e) From the histogram in part (a), the data appears to be approximately normal.

(f) Approximately 68% of the data values should lie between \$22,762 and \$31,492.

(g) Approximately 95% of the data values should lie between \$18,397 and \$35,857.

(h) The actual number of values in the interval given in part (f) is 18. Thus the actual proportion of values that are in the interval is $(18/25) \times 100\% = 72\%$.

(i) The actual number of values in the interval given in part (g) is 24. Thus the actual proportion of values that are in the interval is $(24/25) \times 100\% = 96\%$.

13. (a) A histogram using seven classes is shown below for the personal income per capita for all the data in the table.

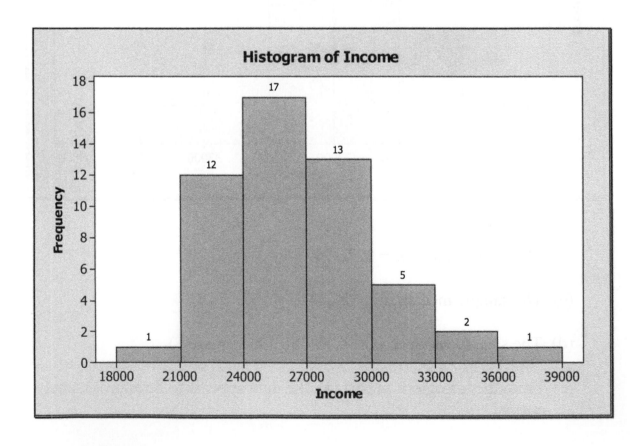

(b) The sample mean $\overline{x}$ = $26,738.

(c) The sample median m = $26,474.

(d) The sample variance is s^2 = 15,152,879.

(e) Yes. From the histogram in part (a), the data appears to be approximately normal.

(f) Approximately 68% of the data values should lie between $22,845 and $30,631.

(g) Approximately 95% of the data values should lie between $18,952 and $34,524.

(h) The actual number of values in the interval given in part (f) is 36. Thus the actual proportion of values that are in the interval is $(36/50)\times100\%$ = 72%.

(i) The actual number of values in the interval given in part (g) is 48. Thus the actual proportion of values that are in the interval is $(48/50)\times100\%$ = 96%.

SECTION 3.7 -- THE SAMPLE CORRELATION COEFFICIENT

PROBLEMS

1. For all three data sets, you can use Program 3-2 from the CD in the text or any other statistical software to show that the absolute value of the correlation coefficient, $|r| = 1$. This implies that the three sets of data have the same degree of correlation. Observe from the fitted line plots for

the three data sets that the regression line passes through all the points with a positive slope. Hence there is a perfect positive association for all three data sets.

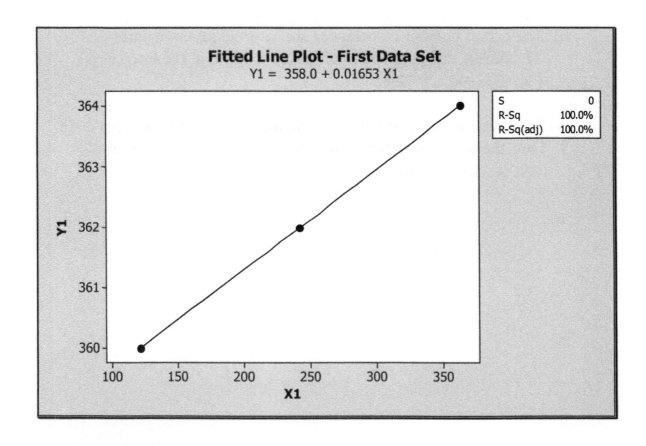

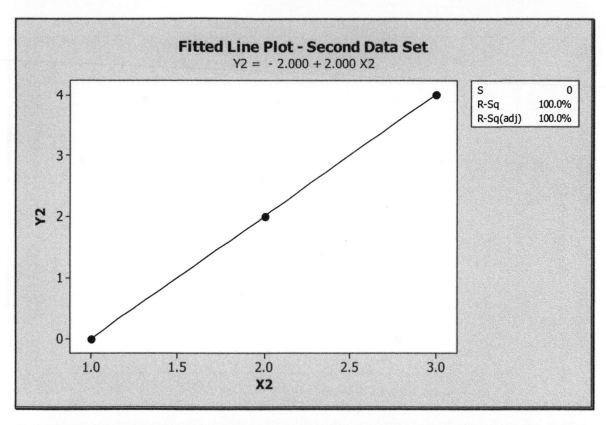

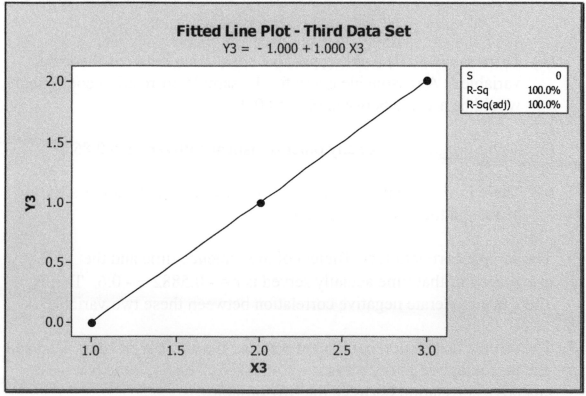

3. (a) A scatter diagram for the Daughter's IQ score versus the Mother's IQ score is shown below.

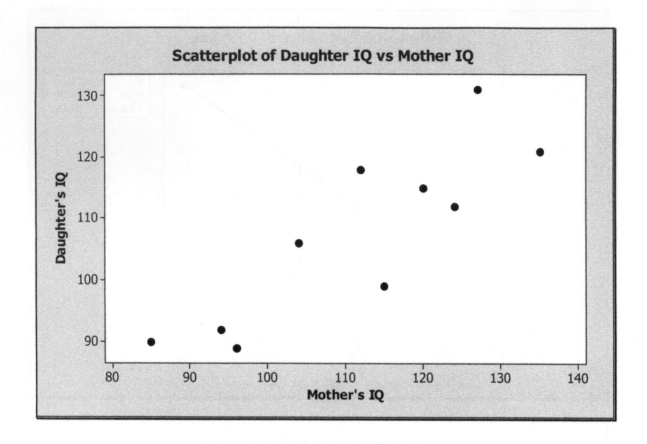

(b) There seems to be a strong positive correlation between these two variables. A reasonable guess for the sample correlation coefficient would be a value between 0.8 and 0.9.

(c) Using Program 3-2 or any other statistical software, r = 0.862.

(d) There is a strong positive correlation between the daughter's IQ score and the mother's IQ score.

5. The sample correlation coefficient of the sentence time and the proportion of that time actually served is r = - 0.5882 ≈ - 0.6. That is, there is a moderate negative correlation between these two variables.

7. The sample correlation coefficient between the number of cases filed and the percentage of guilty pleas is r = - 0.4416. That is, these is a weak negative correlation between the two variables.

9. The sample correlation coefficient between the 1979 and 1985 per capita incomes for the given cities is $r = 0.9460$.

11. Using all of the data: The sample correlation coefficient between death rates of ischemic heart disease and of chronic liver disease is $r = -0.3302$.

    Using the first seven data values: The sample correlation between death rates of ischemic heart disease and of chronic liver disease is $r = -0.0447$.

13. (a) Using all of the data values: The sample correlation coefficient between death rates of lung cancer and of bronchitis, emphysema, and asthma is $r = 0.2530$.
    (b) Using the first seven data values: The sample correlation coefficient between death rates of lung cancer and of bronchitis, emphysema, and asthma is $r = -0.3033$.

15. (a) No. Correlation is not causation.

    (b) No.

    (c) No.

    (d) One possible explanation is that this sample of young adults ate a lot of 'junk' foods while watching TV. In addition, a great deal of TV watching would imply a great deal of inactivity which in turn can
    cause weight gain and hence an increase in cholesterol levels.

17. No. Correlation between having a gun at home for protection and the likelihood of being murdered does not imply causation.

REVIEW PROBLEMS

1. (a) An example of data set that is symmetric about zero and contains 4 distinct values is the set: -2, -1, 1, 2.

   (b) An example of a data set that is symmetric about zero and contains 5 distinct values is the set: -2, -1, 0, 1, 2.

   (c) Part (a): the mean $\bar{x} = 0$; the median m = 0.
   Part (b): the mean $\bar{x} = 0$; the median m = 0.

3. (a) The median for the ages is m = 29.3.

   (b) Yes. The median m = 29.3 is the median of all of the 50 medians. This would imply that 29.3 is the median age of all the people in the U.S.

   (c) The first quartile or the 25-percentile = 27.675.

   The second quartile or the median or the 50-percentile = 29.3.

   The third quartile or the 75-percentile = 31.125.

   (d) First arrange the data in ascending order. Now n×p = 50(0.9) = 45. Thus the sample 90-percentile will be the average of the 45th and 46th values in the ordered data set. That is, the 90-percentile is (31.6 + 31.8)/2 = 31.7.

5. No. This can be seen from the histogram below. Seem more right skewed.

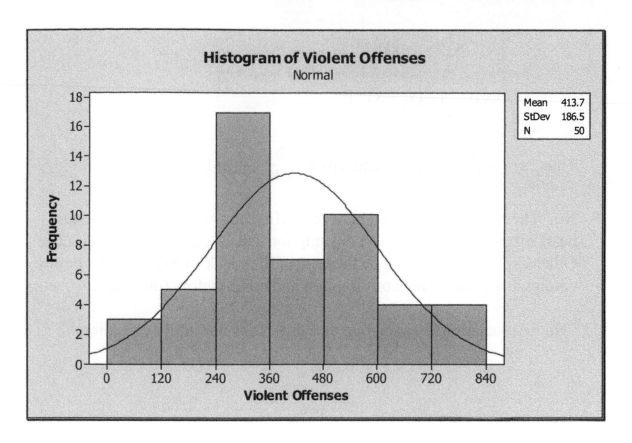

7. Given: $y_i = a + bx_i$, for i = 1, 2, ..., n.

$$r = \frac{\sum\limits_{i}^{n}(x_i - \bar{x})(y_i - \bar{y})}{\sqrt{\sum\limits_{i}^{n}(x_i - \bar{x})^2 \sum\limits_{i}^{n}(y_i - \bar{y})^2}}$$

Also, $\bar{y} = a + b\bar{x}$. Now, substituting in r and simplifying gives

$$r = \frac{\sum\limits_{i}^{n}(x_i - \bar{x})b(x_i - \bar{x})}{\sqrt{\sum\limits_{i}^{n}(x_i - \bar{x})^2 \sum\limits_{i}^{n}[b(x_i - \bar{x})]^2}}$$

which implies that

$$r = \frac{b\sum_{i}^{n}(x_i - \bar{x})^2}{\pm\sqrt{b^2}\sqrt{[\sum_{i}^{n}(x_i - \bar{x})^2]^2}}. \quad \text{That is, } r = \frac{b}{\pm b}.$$

Thus (a) $r = \dfrac{b}{b} = 1$ if $b > 0$ and (b) $r = \dfrac{b}{-b} = -1$ if $b < 0$.

9. No. There is an association between coffee consumption and coronary heart disease not that coffee consumption causes coronary heart disease. Other explanations would be the type of diet, lack of exercise etc. or any combinations of coffee consumption and other factors (diet etc.).

11. No. Association between two variables does not imply causation.

Chapter 4   PROBABILITY

SECTION 4.2 -- SAMPLE SPACE AND EVENTS OF AN EXPERIMENT

PROBLEMS

1. This experiment is with replacement.

   Let the event of a red ball be defined as R.

   Let the event of a blue ball be defined by B.

   Let the event of a yellow ball be defined by Y.

   (a) The sample space S for this experiment is

   $$S = \{(R, R), (R, B), (R, Y), (B, B), (B, R), (B, Y), (Y, Y),$$
   $$(Y, R), Y, B)\}.$$

   (b) Let A be the event that the first ball drawn is yellow. Then

   $$A = \{(Y, Y), (Y, B), (Y, R)\}.$$

   (c) Let B be the event that the same ball is drawn twice. Then

   $$B = \{(R, R), (B, B), (Y, Y)\}.$$

3. Let UM be the event of choosing University of Michigan; RC the event of choosing Reed College; SJSC the event of choosing San Jose State College; YU the event of Yale University; OSU the event of Oregon State University.

   Since the outcomes of the experiment are the colleges that Audrey and

Charles choose to attend, then

(a) The sample space

$$S = \{(UM, OSU), (UM, SJSC), (RC, OSU), (RC, SJSC),$$
$$(SJSC, OSU), (SJSC, SJSC), (YU, OSU), (YU, SJSC),$$
$$(OSU, OSU), (OSU, SJSC)\}.$$

(b) Let A be the event that Audrey and Charles attend the same school, then

$$A = \{(SJSC, SJSC), (OSU, OSU)\}.$$

(c) Let B be the event that Audrey and Charles attend different schools, then

$$B = \{(UM, OSU), (UM, SJSC), (RC, OSU), (RC, SJSC),$$
$$(SJSC, OSU), (YU, OSU), (YU, SJSC), (OSU, SJSC)\}.$$

(d) Let C be the event that Audrey and Charles attend schools in the same state, then

$$C = \{(RC, OSU), (SJSC, SJSC), (OSU, OSU)\}.$$

5. Since the outcomes for this experiment are location and mode of travel, then the sample space

$$S = \{(France, fly), (France, boat), (Canada, drive), (Canada, train),$$
$$(Canada, fly)\}.$$

Since A is the event that the family flies to their destination, then

$$A = \{(France, fly), (Canada, fly)\}.$$

7. $S = \{1, 2, 3, 4, 5, 6\}$; $A = \{1, 3, 5\}$; $B = \{4, 6\}$; $C = \{1, 4\}$.

(a) $A \cap B = \{1, 3, 5\} \cap \{4, 6\} = \varnothing$ (null set).

(b) $B \cup C = \{4, 6\} \cup \{1, 4\} = \{1, 4, 6\}$.

(c) $A \cup (B \cap C) = \{1, 3, 5\} \cup (\{4, 6\} \cap \{1,4\}) = \{1, 3, 4, 5\}$.

(d) $(A \cup B)^c = (\{1,3, 5\} \cup \{4, 6\})^c = \{2\}$.

9. (a) The sample space for this experiment

$$S = \{(1, g), (1, f), (1, s), (1, c), (0, g), (0, f), (0, s), (0, c)\}.$$

(b) Let A be the event that a patient is in serious or critical condition and having no insurance, then

$$A = \{(0, s), (0, c)\}.$$

(c) Let B be the event that the patient is in good or fair condition, then

$$B = \{(1, g), (1, f), (0, g), (0, f)\}.$$

(d) Let C be the event that the patient have insurance, then

$$C = \{(1, g), (1, f), (1, s), (1, c)\}.$$

11. (a) A^c is the event that the die lands on an odd number.

(b) $(A^c)^c$ is the event that the die lands on an even number.

(c) In general, $(A^c)^c = A$. That is, the complement of the complement of an event is the event itself.

13. (a)   (b)

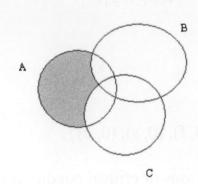

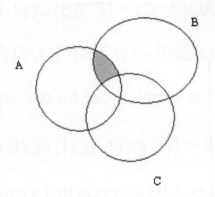

(c)

(d)

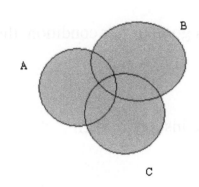

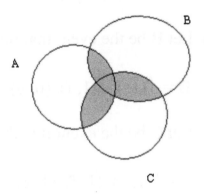

(e)

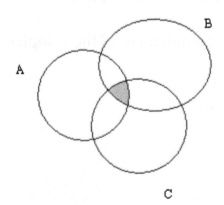

SECTION 4.3 -- PROPERTIES OF PROBABILITY

PROBLEMS

1. Note the sample points in the events E are disjoint
   (mutually exclusive). This is also true for events F and G.

   (a) $P(E) = P\{1, 3, 5\} = P(1) + P(3) + P(5) = 0.1 + 0.15 + 0.1 = 0.35.$

   $P(F) = P\{2, 4, 6\} = P(2) + P(4) + P(6) = 0.2 + 0.15 + 0.3 = 0.65.$

   $P(G) = P\{1, 4, 6\} = P(1) + P(4) + P(6) = 0.1 + 0.15 + 0.3 = 0.55.$

   (b) $P(E \cup F) = P\{1, 2, 3, 4, 5, 6\} = P(S) = 1.$

   (c) $P(E \cup G) = P\{1, 3, 4, 5, 6\} = 1 - P\{2\} = 1 - 0.2 = 0.8.$

   (d) $P(F \cup G) = P\{1, 2, 4, 6\} = P(1) + P(2) + P(4) + P(6)$
   $$= 0.1 + 0.2 + 0.15 + 0.3 = 0.75.$$

   (e) $P(E \cup F \cup G) = P(S) = 1.$

   (f) $P(E \cap F) = P(\text{null set}) = 1 - P(S) = 1 - 1 = 0.$

   (g) $P(F \cap G) = P\{4, 6\} = P(4) + P(6) = 0.15 + 0.3 = 0.45.$

   (h) $P(E \cap G) = P\{1\} = 0.1.$

   (i) $P(E \cap F \cap G) = P(\text{null set}) = 0.$

3. Let A be the event that the next child born in a Houston hospital has
   phenylketonuria. Since 1 in every 10,000 live birth in the U.S has
   phenylketonuria, then,
   $$P(A) = \frac{1}{10,000} = 0.00001.$$

5. If A and B are disjoint, then P(A) + P(B) = 1.2 is impossible since the sum
   cannot exceed 1 because there is no overlapping common portion to the events A and B.

   If A and B are not disjoint, then P(A) + P(B) = 1.2 is possible because there will be an overlapping common portion to the events A and B.

7. (a) Let A be the event of selling the entire stock of 1100 trees.

   The demand in this case has to lie between 1100 and 2000 trees in order for the entire stock of 1100 to be sold. Thus,

   P(A) = P(demand is 1100 or demand is 1400 or demand is 1600 or demand is 2000).

   Since the events are disjoint, then

   P(A) = P(demand is 1100) + P(demand is 1400)

   + P(demand is 1600) + P(demand is 2000)

   = 0.2 + 0.3 + 0.4 + 0.1 = 1.

   (b) Let B be the event of selling the entire stock of 1400 trees.

   The demand in this case has to lie between 1400 and 2000 in order for the entire stock of 1400to be sold. Thus,

   P(B) = P(demand is 1400 or demand is 1600 or demand is 2000).

   Since the events are disjoint, then

   P(B) = P(demand is 1400) + P(demand is 1600)

   + P(demand is 2000)

   = 0.3 + 0.4 + 0.1 = 0.8.

(c) Let C be the event of selling the entire stock of 1600 trees.

The demand in this case has to lie between 1600 and 2000 in order for the entire stock to be sold. Thus,

P(B) = P(demand is 1600 or demand is 2000).

Since the events are disjoint, then

P(B) = P(demand is 1600) + P(demand is 2000)

= 0.4 + 0.1 = 0.5.

(d) Let D be the event of selling the entire stock of 2000 trees.

The demand in this case has to be 2000 in order for the entire stock of 2000 to be sold. Thus,

P(B) = P(demand is 2000) = 0.1.

9. (a) P(3 or fewer errors) = 1 - P(4 or more errors) = 1 - 0.05 = 0.95.

(b) P(2 or fewer errors) = P(0 error) + P(1 error) + P(2 errors)

= 0.2 + 0.35 + 0.25 = 0.8.

(c) P(no errors) = P(0 error) = 0.20.

11. Let A be the event of rain. Let B be the event of cloudiness. Then,

P(A) = 0.4, P(B) = 0.5, and P(A and B) = P(A∩B) = 0.2. Thus,

P(A or B) = P(A∪B) = P(A) + P(B) - P(∩B)

= 0.4 + 0.5 - 0.2 = 0.7.

13. Let A be the event that a person in the U.S. is obese.

    Let B be the event that a person in the U.S. suffer from diabetes.

    $P(A) = 0.30$, $P(B) = 0.03$, $P(A \cap B) = 0.02$, then

    $P(A \cup B) = P(A) + P(B) - P(A \cap B)$

    $\qquad\qquad = 0.3 + 0.03 - 0.02 = 0.31$.

15. Let A be the event of purchasing a suit.

    Let B be the event of purchasing a tie.

    $P(A) = 0.3$, $P(B) = 0.2$, $P(A \cap B) = 0.1$.

    Now the probability of purchasing at least one of the items is

    $P(A \cup B) = P(A) + P(B) - P(A \cap B) = 0.3 + 0.2 - 0.1 = 0.4$.

    Thus P(purchasing neither a suit nor a tie) $= 1 - P(A \cup B)$

    $\qquad = 1 - 0.4 = 0.6$.

17. (a) $B^c \cap A$

    (b) $A \cap B$

    (c) $A^c \cap B$

    (d) $P(A \cup B) = P(I) + P(II) + P(III)$

    (e) $P(A) = P(I) + P(II)$

    (f) $P(B) = P(II) + P(III)$

    (g) $P(A \cap B) = P(II)$

(h) From part (d), $P(A \cup B) = P(I) + P(II) + P(III)$ and from part (e),

P(I) = P(A) - P(II). Substituting for P(I), we get

$P(A \cup B) = P(A) - P(II) + P(II) + P(III) = P(A) + P(III)$.

Now from part (f) $P(III) = P(B) - P(II)$, so

$P(A \cup B) = P(A) + P(B) - P(II) = P(A) + P(B) - P(A \cap B)$

since $P(II) = P(A \cap B)$ from the diagram.

SECTION 4.4 -- EXPERIMENTS HAVING EQUALLY LIKELY OUTCOMES

PROBLEMS

1. Let A be the event that a student did not sleep through the alarm.

   Now, the number of students that did not sleep through the alarm

   $$= 216 - 128 = 88.$$

   Since the number of students in that particular dorm is 216, then

   $$P(A) = \frac{88}{216} = 0.4074.$$

3. (a) Let A be the event that the selected card is an ace. Then,

   $$P(A) = \frac{4}{52} = 0.0769.$$

   (b) P(selected card is not an ace $= 1 - P(A) = 1 - 0.0769 = 0.9231$.

   (c) Let B be the event that the selected card is a spade. Then,

   $$P(B) = \frac{13}{52} = 0.25.$$

   (d) Let C be the event that the selected card is the ace of spade. Then,

   $$P(C) = \frac{1}{52} = 0.0192.$$

5. (a) P(car chosen is from the U.S.) = (5,027,425)/(12,328,305 + 10,239,949 + ... + 3,147,584)
   = 0.1443.

(b) P(car chosen is from South Korea) = (3,147,584)/(12,328,305 +

10,239,949 + ... + 3,147,584)

= 0.0761.

7. Let A be the event that a person is sign up for the swimming class.

Let B be the event that a person is sign up for a calisthenics class.

Given: $P(A \cup B) = 44/100 = 0.44$; $P(A) = 26/100 = 0.26$;

$P(B) = 28/100 = 0.28$.

(a) P(patient is not in the exercise program) = $[P(A \cup B)]^c = 1 - P(A \cup B)$

= 1 - 0.44 = 0.56.

(b) Recall, $P(A \cup B) = P(A) + P(B) - P(A \cap B)$. Thus,

$0.44 = 0.26 + 0.28 - P(A \cap B)$ or $P(A \cap B) = 0.26 + 0.28 - 0.44 = 0.1$.

That is, P(person is enrolled in both classes) = 0.1.

9. Let A be the event that the student wears a ring.

Let B be the event that the student wears a necklace.

Given: $P(A) = 0.20$; $P(B) = 0.3$; $[P(A \cup B)]^c = 0.6$.

(a) $P(A \cup B) = 1 - [P(A \cup B)]^c = 1 - 0.6 = 0.4$.

(b) Recall, $P(A \cup B) = P(A) + P(B) - P(A \cap B)$, thus

$0.4 = 0.2 + 0.3 - P(A \cap B)$ and so $P(A \cap B) = 0.2 + 0.3 - 0.4 = 0.1$.

That is, P(student is wearing a ring and a necklace) = 0.1.

11. Let A be the event that a club member plays tennis.

Let B be the event that a club member plays squash.

Given: $P(A) = 44/120$; $P(B) = 30/120$; $P(A \cap B) = 18/120$.

P(member plays neither tennis nor squash) = 1 - $P(A \cup B)$.

Now, $P(A \cup B) = P(A) + P(B) - P(A \cap B)$, so

$P(A \cup B) = P(A) + P(B) - P(A \cap B) = 44/120 + 30/120 - 18/120$

$= 56/120 = 0.4667$. Thus, the number of members that play either tennis

or squash = $(56/120) \times 120 = 56$.

13. First choose a person. Now, from the remaining 19 people, the probability that the next person chosen will be married to the first person chosen will be 1/19.

15. (a) P(first key selected opens the door) = $1/10 = 0.1$.

(b) P(all ten keys are tried)

$= (9/10) \times (8/9) \times (7/8) \times (6/7)) \times (5/6) \times (4/5)) \times (3/4) \times) \times (2/3) \times (1/2)\ 0.1$.

17. (a) P(rain on January 5) = 10/31.

(b) P(rain August 12) = 9/31.

(c) P(rain April 15) = 10/30 = 1/3.

(d) P(rain May 15) = 11/31.

(e) P(rain October 12) = 7/31.

SECTION 4.5 -- CONDITIONAL PROBABILITY AND INDEPENDENCE

PROBLEMS

1. Let A be the event that an adult is obese.

   Let B be the event that an adult suffer from diabetes.

   Given: $P(A) = 0.3$; $P(B) = 0.03$; $P(A \cap B) = 0.02$.

   (a) $P(B|A) = [P(A \cap B)]/P(A) = 0.02/0.3 = 0.0667$.

   (b) $P(A|B) = [P(A \cap B)]/P(B) = 0.02/0.03 = 0.6667$.

3. (a) Let A be the event of a woman.

   Let B be the event of person earning over $25,000.

   We need to find $P(A|B) = [P(A \cap B)]/P(B)$

   Since $25,000 is the class mark for the interval $20,000 to $30,000, one way to solve this problem is to assume all values in the interval is equal to the class mark.

   Thus, $P(A|B) = [P(A \cap B)]/P(B) = (6{,}291 + 6{,}555 + 5{,}169 + 8{,}255 + 947)/(6{,}291 + 6{,}555 + 5{,}169 + 8{,}255 + 947 + 5{,}081 + 6{,}386 + 6{,}648 + 20{,}984 + 7{,}377) = 27{,}217/73{,}693 = 0.3693$.

   (b) $P(B|A) = [P(A \cap B)]/P(A) = (6{,}291 + 6{,}555 + 5{,}169 + 8{,}255 + 947)/(31{,}340) = 27{,}217/31{,}340 = 0.8684$.

5. (a) P(resident is less than 10 years) $= 4{,}200/28{,}900 = 0.1453$.

   (b) P(resident between 10 and 20 years old) $= 5{,}100/28{,}900 = 0.1765$.

   (c) P(resident is between 20 and 30 years old) $= 6{,}200/28{,}900 = 0.2145$.

(d) P(resident is between 30 and 40 years old) = 4,400/28,900 = 0.1522.

7. Let A be the event that a club member plays chess.

Let B be the event that a club member plays bridge.

Given: P(A) = 40/120; P(B) = 56/120; P(A∩B) = 26/120.

(a) P(A|B) = [P(A∩B)]/P(B) = (26/120)/(56/120) = 0.4643.

(b) P(B|A) = [P(A∩B)]/P(A) = (26/120)/(40/120) = 0.65.

9. (a) P(going to graduate school | no business or teaching)
       = 0.262/0.689 = 0.3803.

(b) P(going into teaching or graduate school | no business or teaching)
       = 0.262/0.689 = 0.3803.

(c) P(going into communications or graduate school | no business or
       teaching) = 0.346/0.689 = 0.5022.

(d) P(not going into science/technology | no business or teaching)
       = 0.609/0.689 = 0.8839.

(e) P(not going into communications or business | no business or
teaching)
       = 0.605/0.689 = 0.8781.

(f) P(not going into science/technology or government/politics | no
       business or teaching) = 0.526/0.689 0.7634.
****

11. Let A be the event that the first card drawn is an ace.

Let B be the event that the second card drawn is an ace.

*NOTE: For aces to be of different suits we assume selection is
        without replacement.*

Now $P(A \cap B) = P(B|A)P(A) = (3/51)(4/52) = 0.0045.$

****

13. Let A be the event of a psychiatrist being chosen.

Let B be the event of a psychologist being chosen.

The sample space is S = {AA, AB, BA, BB}.

P(at least one psychologist being chosen)

$= 1 - P(AA) = 1 - (30/54)(29/53) = 0.6960.$

15. Let S be the event of a spade. Let C be the event of a club. Let D be the event of a diamond. Let H be the event of a heart.

(a) The event of neither card being a spade corresponds to the

following event E = {CC, CD, CH, DC, DD, DH, HC, HD, HH}.

$P(CC) = P(DD) = P(HH) = (13/52)(12/51)$ and

$P(CD) = P(CH) = P(DC) = P(DH) = P(HC) = P(HD)$

$= (13/52)(13/51).$ Thus, $P(E) = 3(13/52)(12/51) + 6(13/52)(13/51)$

$= 0.5588.$

(b) P(at least one is a spade) $= 1 - P(E) = 1 - 0.5588 = 0.4412.$

(c) P(both are spades) $= P(SS) = (13/52)(12/51) = 0.0588.$

17. Given $P(B|A) > P(B).$ Now, $P(B|A) = [P(A \cap B)]/P(A)$ which implies

that $[P(A \cap B)]/P(A) > P(B)$ or $P(A \cap B) > P(A) \times P(B)$ since $P(A) > 0.$

Now $P(A|B) = [P(A \cap B)]/P(B)$ or $P(A \cap B) = P(A|B) \times P(B).$ Substituting

for $P(A \cap B)$ gives $P(A|B) \times P(B) > P(A) \times P(B)$ or $P(A|B) > P(A).$

19. Let A be the event of a new branch office in Chicago.

    Let B be the event of Norris being named the manager.

    Given: $P(A) = 0.40$; $P(B|A) = 0.60$. Thus,

    P(new branch and Norris being the manager) = $P(A \cap B)$

    = $P(B|A) \times P(A) = (0.6)(0.4) = 0.24$.

21. Let A be the event that the patient survive on the operating table.

    Let B be the event that the patient survives from the after effects.

    Given: $P(A) = 0.8$; $P(B|A) = 0.85$. So,

    P(patient surviving the operation and the after effects) = $P(A \cap B)$

    = $P(B|A) \times P(A) = (0.8)(0.85) = 0.68$.

23. Let A be the event that a club member plays chess.

    Let B be the event that a club member plays bridge.

    Given: $P(A) = 40/120$; $P(B) = 56/120$; $P(A \cap B) = 26/120$.

    (a) $P(A \cup B) = P(A) + P(B) - P(A \cap B) = 70/120 = 0.5833$.

    (b) $120 - [(40 - 26) + 26 + (56 - 26)] = 50$.

    (c) P(both play chess) = $(40/120)(39/119) = 0.1092$.

    (d) P(neither one plays chess or bridge) = $(50/120)(49/119) = 0.1716$.

    (e) P(both play either chess or bridge) = $1 - 0.1716 = 0.8284$.

25. Let A be the event of John hitting the duck.

Let B be the event of Jim hitting the duck.

Given: $P(A) = 0.3$; $P(B) = 0.1$.

(a) P(A | exactly one shot hits the duck)

= P(A and exactly one shot hits the duck)/P(exactly one shot hits the duck)

Now P(exactly one shot hits the duck) = $P(A \cap B^C) + P(A^C \cap B)$

= (0.3)(0.9) + (0.7)(0.1) = 0.34, and

P(A and exactly one shot hits the duck) = $P(A \cap B^C)$ = (0.3)(0.9).

So, P(A | exactly one shot hits the duck) = (.3)(0.9)/0.34 = 0.7941.

Also, P(B | exactly one shot hits the duck) = (0.7)(0.1)/0.34 = 0.2059.

(b) P(A | duck was hit) = P(A and duck was hit)/P(duck was hit)

Now, P(duck was hit) = $P(A) + P(B) - P(A \cap B)$

= 0.3 + 0.1 - (0.3)(0.1) = 0.37 (because of independence)

So, P(A | duck was hit) = P(A and duck was hit)?P(duck was hit)

= 0.3/0.37 = 0.8108.

Also, P(B | duck was hit) = P(B and duck was hit)?P(duck was hit)

= 0.1/0.37 = 0.2703.

27. (a) P(stock at original price after 2 days) =

P[(up 1st and down 2nd day) or (down 1st and up 2nd day)] =

P(up 1st and down 2nd day) + P(down 1st and up 2nd day) =

P(up 1st) P(down 2nd) + P(down on 1st) P(up on 2nd)

$= (1/2)(1/2) + (1/2)(1/2) = 1/2.$

(b) P[(up and down and up) or (down and up and up) or (up and up and down)]

$= 3(1/2)(1/2)(1/2) = 3/8.$

(c) P(up on 1st day | price increased by one unit after 3 days)

= P(up on 1st day and the price increased by one unit after 3 days)/
    P(the price increased by one unit after 3 days)

$= (2/8)/(3/8) = 2/3.$

29. Let H be the event of a head and T be the event of a tail.

For a tail on the 5th toss implies successive heads on the 1st four tosses.

Thus, P(coin has to be tossed at least 5 times) = P(HHHH) = $(1/2)^4$

$= 1/16.$

31. No. If the friends do not know each other, then one can assume that they would not call each other. Thus, the busy signals will be independent of each other.

33. Let A be the event of selecting an ace.

Let B be the event of selecting a spade.

P(A) = 4/52 = 1/13; P(B) = 13/52 = 1/4.

Now, P(A|B) = [P(A∩B)]/P(A) = P(ace of spade)/P(B)

$= (1/52)/(13/52) = 1/13 = P(A)$. Hence A and B are independent.

35. Let A be the event of the first person's birthday.

    Let B be the event of the second person's birthday which is different from the first person's.

    So, $P(A) = 365/365 = 1$; $P(B) = 364/365$. Thus the probability of different birthdays is $P(A \cap B) = (1)(364/365)$. Thus, the probability of having the same birthday will be $1 - 364/365 = 1/365$.

37. (a) P(current flowing between A and B) = P(both relays are closed)

    $= (0.8)(0.8) = 0.64$.

    (b) Let C be the event that the upper relay is closed.

    Let D be the event that the lower relay is closed.

    $P(C) = P(D) = 0.8$. Now, P(current flowing between A and B)

    is $P(C \cup D) = P(C) + P(D) - P(C \cap D) = 0.8 + 0.8 - (0.8)(0.8) = 0.96$.

    (c) Let E be the event that both upper relays are closed.

    Let F be the event that both lower relays are closed.

    $P(E) = P(F) = (0.8)(0.8) = 0.64$. Now,

    P(current flowing between A and B) = $P(E \cup F) =$

    $P(E) + P(F) - P(E \cap F) = 0.64 + 0.64 - (0.64)(0.64) = 0.8704$.

39. Let A be the event that the first ball drawn is white.

    Let B be the event that the second ball drawn is black.

The sample space is S = {AA, AB, BA, BB}.

Assume balls are drawn with replacement.

P(A|B) = [P(A∩B)]/P(B) = P(AB)/P(B). Now P(AB) = (1/2)(1/2) and

P(B) = P(AB or BB) = (1/2)(1/2) + (1/2)(1/2), so P(A|B) = 1/2. Also,

P(A) = P(AA or AB) = (1/2)(1/2) + (1/2)(1/2) = 1/2. Since

P(A|B) = P(A), then A and B are independent if the first ball is replaced before the second ball is drawn.

41. (a) P(raining on each trip) = (4/31)(3/30)(16/31) = 0.0067.

    (b) P(dry on all 3 trips) = (27/31)(27/30)(15/31) = 0.3793.

    (c) P(rain in Phoenix and Mobile but not in LA) = (4/31)(16/31)(27/30)

        = 0.0599.

    (d) P(rain in Mobile and LA but not in Phoenix) = (16/31)(3/30)(27/31)

        = 0.0450.

    (e) P(rain in Phoenix and LA but not in Mobile) = (4/31)(3/30)(15/31)

        = 0.0062.

    (f) P(raining in exactly two of the three trips) =

        answer in (c) + answer in (d) + answer in (e) = 0.1111.

43. Let A be the event that the child will receive the CF (cystic fibrosis)

    from the mother and let B be the corresponding event for the father.

Given: $P(A) = P(B) = 0.5$.

(a) P(child developing CF) = $P(A \cap B) = P(A)P(B) = (0.5)(0.5) = 0.25$.

(b) P(sibling carrying the CF gene) =

P(25 year old person receiving the gene from either his mother or father) =

P(25 year old carrying the CF gene | sibling has CF genes) =

P(25 year old carrying the CF gene and sibling having CF genes)/ P(sibling has CF genes).

Now, P(25 year old receiving one CF gene) = 0.5, and

P(sibling has CF genes) = P(25 year old received at least one CF gene) =

P(25 year old receive 2 CF genes) + P(25 year old receive 1 CF gene)

= (0.5)(0.5) + 0.5 = 0.75.

Thus, P(sibling carrying the CF gene) = 0.5/0.75 = 2/3 = 0.6667.

SECTION 4.6 BAYES' THEOREM

PROBLEMS

1. Let A be the event of the fair coin. Let B be the event of the biased coin.

   Let H be the event of a head and T be the event of a tail.

   Given: $P(A) = P(B) = 0.5$; $P(H|A) = 0.5$; $P(H|B) = 0.6$.

   (a) $P(H) = P(A \cap H \text{ or } B \cap H) = P(A \cap H) + P(B \cap H) =$

   $(0.5)(0.5) + (0.5)(0.6) = 0.55$.

   (b) $P(A|T) = P(A \cap T)/P(T) = (0.5)(0.5)/[(0.5)(0.5) + (0.5)(0.4)] = 0.5556$.

3. Let G be the event that the suspect is guilty. Let L is the event of being left handed. Let R be the event of being right handed. Let N be the event of not guilty.

   Given: $P(L|N) = 0.18$; $P(G) = 0.6$. The following tree diagram shows the
   relevant probabilities.

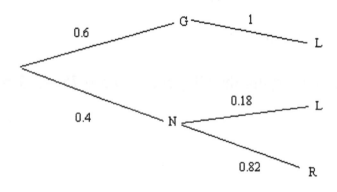

   (a) P(suspect is left handed) $= P(L|G)P(G) + P(L|N)P(N)$

   $= 1 \times (0.6) + (0.4)(0.18) = 0.672$.

   (b) $P(G|L) = [P(L|G)P(G)]/[P(L|G)P(G) + P(L|N)P(N)]$

   $= (0.6)(1)/0.672 = 0.893$.

5. Let A be the event that the person has the disease. Le B be the event that the test was positive.

$P(A|B) = [P(B|A)P(A)]/[P(B|A)P(A) + P(B|A^C)P(A^C)] =$

$[(0.97)(0.02)]/[(0.97)().02) + (0.03)(0.98)] = 0.3975.$

7. Let R be the event of a Republican. Let D be the event of a Democrat. Let E be the event of discontinuing affirmative action city hiring policies.

   Given: $P(R) = 0.52$; $P(D) = 0.48$; $P(E|R) = 0.64$; $P(E|D) = 0.42$.

   (a) $P(E) = P(E|R)P(R) + P(E|D)P(D) = (0.64)(0.52) + (0.42)(0.48)$

   $= 0.5344.$

   (b) $P(R|E) = [P(E|R)P(R)]/[P(E|R)P(R) + P(E|D)P(D)] = 0.6228.$

SECTION 4.7 COUNTING PRINCIPLES

PROBLEMS

1. Number of different 7-place license plates $= (26)^3 \times (10)^4 = 175,760,000$.

3. Since $9! = 362,880$, then $10! = 10 \times 9! = 10 \times 362,880 = 3,628,800$.

5. Number of area codes possible $= 8 \times 2 \times 9 = 144$.

   Number of area codes starting with $4 = 1 \times 2 \times 9 = 18$ since there is only one way to select the first digit of a 4.

7. (a) Number of assignments for the workers $= 4! = 24$.

   (b) If workers 1 and 2 are only qualified for jobs 1 and 2 and workers 3 and 4 are only qualified for jobs 3 and 4, then the number of ways of assigning these workers $= 2(2 \times 1) + 2(2 \times 1) = 8$.

9. $\binom{8}{4} = 70$; $\binom{9}{2} = 36$; $\binom{7}{6} = 7$; $\binom{10}{3} = 120$.

11. (a) Number of possible choices $= \binom{5}{4} = 5$.

   (b) P(both French and Spanish are chosen) $= \dfrac{\binom{2}{2} \times \binom{3}{2}}{\binom{5}{4}} = 3/5 = 0.6$.

13. (a) P(no defective) $= \dfrac{\binom{86}{10} \times \binom{14}{0}}{\binom{100}{10}} = 0.2046$.

   (b) P(one defective) $= \dfrac{\binom{86}{9} \times \binom{14}{1}}{\binom{100}{10}} = 0.3719$.

   (c) P(shipment will be rejected) $=$ P(2 or more defectives)

   $= 1 - \{P(\text{zero defective}) + P(\text{one defective})\}$

$$= 1 - (0.2046 + 0.3719) = 0.4235$$

15. (a) $P(6 \text{ women and } 6 \text{ men}) = \dfrac{\dbinom{22}{6} \times \dbinom{18}{6}}{\dbinom{40}{12}} = 0.2479.$

(b) $P(8 \text{ women and } 4 \text{ men}) = \dfrac{\dbinom{22}{4} \times \dbinom{18}{8}}{\dbinom{40}{12}} = 0.0573.$

(c) $P(\text{at least } 10 \text{ men}) = P(10 \text{ men}) + P(11 \text{ men}) + P(12 \text{ men})$

$$= \dfrac{\dbinom{22}{10} \times \dbinom{18}{2}}{\dbinom{40}{12}} + \dfrac{\dbinom{22}{11} \times \dbinom{18}{1}}{\dbinom{40}{12}} + \dfrac{\dbinom{22}{12} \times \dbinom{18}{0}}{\dbinom{40}{12}}$$

$$= 0.0177 + 0.0023 + 0.0001 = 0.0201.$$

17. (a) $P(\text{correctly answering all } 5 \text{ problems}) = \dfrac{\dbinom{3}{0} \times \dbinom{7}{5}}{\dbinom{10}{5}} = 0.0833.$

(b) $P(\text{correctly answering at least } 4 \text{ correctly}) = \dfrac{\dbinom{3}{1} \times \dbinom{7}{4}}{\dbinom{10}{5}} + \dfrac{\dbinom{3}{0} \times \dbinom{7}{5}}{\dbinom{10}{5}}$

$$= 0.4167 + 0.0833 = 0.5.$$

19. $P(\text{path goes through the point circled on the grid}) = \dfrac{\dbinom{4}{2}}{\dbinom{7}{3}} = \dfrac{\dbinom{4}{2}}{\dbinom{7}{4}} = 6/35$

$$= 0.1714.$$

REVIEW PROBLEMS

1. (a) P(1st bottle is good) = 9/12 = 3/4.

   (b) P(2nd bottle is good) = P(2nd is good and 1st is good or 2nd is good and 1st is bad)

= P(2nd is good|1st is good)P(1st is good)

   + P(2nd is good|1st is bad)P(1st is bad)

= (8/11)(3/4) + (9/11)(1/4) = 3/4.

(c) P(both bottles are good) = P(1st is good and the 2nd is good)

= (9/12)(8/11) = 6/11.

(d) P(both bottles are bad) = P(1st is bad and the 2nd is bad)

= (3/12)(2/11) = 1/22.

(e) P(one is good and one is bad) = P(1st is good and the 2nd is bad)
                                              + P(1st is bad and the 2nd is good)
= (9/12)(3/11) + (3/12)(9/11) = 9/22.

3. Let S1 be the event of success on the 1st foul shot. Let S2 be the event
    of success on the 2nd foul shot. Let F1 be the event of failure on the 1st
    foul shot. Let F2 be the event of failure on the 2nd foul shot.

    Given: P(S1) = 0.8; P(F1) = 0.2; P(S2|S1) = 0.85; P(S2|F1) = 0.7;

         P(F2|S1) = 0.15; P(F2|F1) = 0.3.

    (a) P(she makes both foul shots) = $P(S1 \cap S2)$ = P(S2|S1)P(S1)

= (0.85)(0.8) = 0.68.

    (b) P(she misses both shots) = $P(F1 \cap F2)$ = P(F2|F1)P(F1)

= (0.3)(0.2) = 0.06.

    (c) P(she makes the 1st but misses the 2nd) = $P(S1 \cap F2)$

= P(F2|S1)P(S1) = (0.15)(0.8) = 0.12.

5. Let B be the event of a boy on the 1st day. Let G be the event of a girl on the 2nd day.

   Given: 13 girls and 11 boys.

   (a) P(B) = 11/24.

   (b) P(G|B) = 13/23.

7. (a) $P(HHHHHH) = (1/2)^6 = 1/64.$ (b) $P(HTHTHT) = (1/2)^6 = 1/64.$

   (c) $P(TTHTH) = (1/2)^6 = 1/64.$

9. Let C be the event of choosing chicken, RB the event of roast beef, R the event of rice, P the event of potatoes, M the event of melon, IC the event of ice cream, and G the event of gelatin.

   (a) S = {(C, R, M), (C, R, IC), (C, R, G), (C, P, M), (C, P, IC), (C, P, G), (RB, R, M), (RB, R, IC), (RB, R, G), (RB, P, M), (RB, P, IC), (RB, P, G)}.

   (b) Let E be the event of a choice that agrees with the person. Then,

   E = {(C, P, IC), (C, P, G), (RB, P, IC), (RB, P, G)}.

   (c) P(apple pie) = 0.

   (d) P(C and R and M) = 1/12.

11. Assume sampling with replacement.

   (a) P(1st key is successful) = 1/3.

   (b) P(2nd key is successful) = 1/3.

   (c) P(3rd key is successful) = 1/3.

   (d) P(2nd key works|1st key did not work) =

P(2nd key works and 1st key did not work)/P(1st key did not work)

$= (1/3)/(2/3) = 1/2.$

13. Let G1 be the event of the 1st truck having good brakes. Let G2 be the event of the 2nd truck having good brakes.

    We need $P(G1 \cap G2) = P(G2|G1)P(G1) = (7/11)(2/3) = 14/33 = 0.4242.$

15. (a) P(ace of spades) = 1/52.

    (b) $P(A \cap B) = P(A)P(B|A) = (51/52)(1/51) = 1/52.$

    (c) Equally.   (d) 1/52.

17. Let G1 be the event of a good 1st year. Let G2 be the event of a good 2nd year. Let B1 be the event of a bad 1st year. Let B2 be the event of a bad 2nd year.

    Given: $P(G2|G1) = 0.2;$ $P(G2|B1) = 0.4;$ $P(G1) = 0.6;$ $P(B2|G1) = 0.8;$

    $P(B2|B1) = 0.6;$ $P(B1) = 0.4.$

    (a) $P(G1 \cap G2) = P(G2|G1)P(G1) = (0.2)(0.6) = 0.12.$

    (b) $P(G1 \cap B2) = P(B2|G1)P(G1) = (0.8)(0.6) = 0.48.$

    (c) $P(B1 \cap B2) = P(B2|B1)P(B1) = (0.6)(0.4) = 0.24.$

    (d) $P(G2) = P(B1 \cap G2$ or $G1 \cap G2) =$

    $P(G2|B1)P(B1) + P(G2|G1)P(G1) = (0.4)(0.4) + (0.2)(0.6) = 0.28.$

    (e) $P(G1|G2) = [P(G1 \cap G2)]/P(G2) = [P(G2|G1)P(G1)]/P(G2)$

    $= [(0.2)(0.6)]/(0.28) = 0.4286.$

19. Let A be the event that a person is above his or her ideal weight. Let B be the event that a person has high blood pressure.

Given: $P(A) = 0.55$; $P(B) = 0.20$; $P(A \cup B) = 0.6$.

Now $P(A \cup B) = P(A) + P(B) - P(A \cap B)$, so $PA \cap B) = 0.15$.

Since $P(A|B) = P(A \cap B)/P(B) = 0.15/0.2 = 0.75$ and $P(A) = 0.55$, then A and B are not independent.

21. (a) P(husband earns less than \$75,000) = 248/500 = 0.496.

    (b) P(wife earns more than \$75,000 | husband earns more than \$75,000)

        = (54/500)/(252/500) = 54/252 = 0.2143.

    (c) P(wife earns more than \$75,000 | husband earns less than \$75,000)

        = (36/500)/(248/500) = 36/248 = 0.1452.

    (d) P(wife earns more than \$75,000) = 90/500 = 0.18 which is not equal to the conditional probabilities in (b) or (c). Hence the salaries of the husband and wife are not independent.

23. Let A be the event of purchasing oil. Let B be the event of purchasing gas.

    Given: $P(A) = 0.08$; $P(B) = 0.86$; $P(A \cup B) = 0.9$.

    (a) $P(A \cup B) = P(A) + P(B) - P(A \cap B)$ from which $P(A \cap B) = 0.04$.

    (b) $P(A|B) = [P(A \cap B)]/P(B) = 0.04/0.86 = 0.0465$.

    (c) $P(B|A) = [P(A \cap B)]/P(A) = 0.04/0.08 = 0.5$.

25. (a) P(exactly one attended the ballet)

        = (0.03)(0.95) + (0.97)(0.05) = 0.077.

    (b) P(at least one attended an opera) = 1 - P(neither attended an opera)

$= 1 - (0.98)(0.97) = 0.0494.$

(c) P(both attended a musical play) $= (0.15)(0.19) = 0.0285.$

Chapter 5 DISCRETE RANDOM VARIABLES

SECTION 5.2 – RANDOM VARIABLES

PROBLEMS

1. The sample space S = {(b, b, b), (b, b, g), (b, g, b), (b, g, g), (g, b, b), (g, b, g), (g, g, b), (g, g, g)}

   Since Y =1 if the family has at least one child of each sex, and Y = 0 otherwise, then there are six points in S associated with Y = 1 and two points associated with Y = 0. Thus,

   P{Y = 0} = 2/8 = ¼ and P{Y = 1} = 6/8 = ¾.

3. Let X = number of tornadoes in a selected year.

   (a) P{X > 900} = 5/12.

   (b) P{X ≤ 800} = 5/12.

   (c) P{X = 852} = 0.

   (d) P{700 < X < 850} = 3/12 = 1/4.

5. Let Y = smaller of the two numbers on the faces of a pair of dice. The possible values of Y are 1, 2, 3, 4, 5, 6. The probability distribution for Y is shown below.

i	1	2	3	4	5	6
P{Y = i}	11/36	9/36	7/36	5/36	3/36	1/36

7. Let X = number of games played.

(a) The possible values of X are 2, 3.

(b) Let H be the home team and V the visiting team. The possible outcomes if the home team wins the series will be HH, HVH, VHH, and if the visiting team wins, the possible outcomes will be VV, VHV, HVV. Thus,

$$P\{X = 2\} = P\{HH \text{ or } VV\} = P\{HH\} + P\{VV\}$$

$$= (0.7)(0.7) + (0.3)(0.3) = 0.58.$$

$$P\{X = 3\} = 1 - 0.58 = 0.42.$$

9. Let X = number of defectives. The possible values of X are 0, 1, 2. The probability distribution for X is shown below.

i	0	1	2
$P\{X = i\}$	1199/1428	55/357	3/476

Note: You can compute these probabilities by following this logic – for instance, to compute $P\{X = 0\} = P\{2 \text{ nondefective}\} =$

$(110/120)(109/119) = (11/12)(109/119) = 1199/1428$ etc.

11. Let X = number of foul shots made. The possible values of X are 0, 1, 2.

The probability distribution for X is shown below.

i	0	1	2
$P\{X = i\}$	0.075	0.325	0.6

Note: You can compute these probabilities by following this logic – for instance, to compute $P\{X = 0\} = P\{\text{missing the 1}^{st} \text{ and missing the 2}^{nd}\}$ $= (0.25)(0.3) = 0.075$, etc.

13. No. $P\{X = 4\} = -1$. Probabilities cannot be negative.

15. Let X = number of children in the household. The possible values of X are 1, 2, 3, 4, 5. The probability distribution for X is shown below.

i	0	1	2	3	4	5
P{X = i}	38/223	82/223	57/223	34/223	10/223	2/223

17. Given: X assumes one of the values 1, 2, 3, 4, or 5; P{X < 3} = 0.4; P{X > 3} = 0.5.

    (a) P{X = 3} = 1 − [P{X > 3} + P{ X < 3}] = 0.1.

    (b) P{X < 4} = P{X ≤ 3} = P{X < 3} + P{X = 3} = 0.4 + 0.1 = 0.5.

19. Let X = number of unsold cakes. The possible values of X are 0, 1, 2, 3. The probability distribution for X is shown below.

i	0	1	2	3
P{X = i}	0.3	0.35	0.2	0.15

SECTION 5.3 – EXPECTED VALUES

PROBLEMS

1. (a) $E[X] = (1)(1/3) + (2)(1/3) + (3)(1/3) = 2$.

    (b) $E[X] = (1)(1/2) + (2)(1/3) + (3)(1/6) = 5/3$.

    (c) $E[X] = (1)(1/6) + (2)(1/3) + (3)(1/2) = 7/3$.

3. $E[\text{Profit}] = (30)(0.4) – (6)(0.6) = 8.4$ 0r $8.40.

5. Since $P\{X = 1\} + P\{X = 2\} + P\{X = 3\} = 1$ and since $P\{X = 1\} = 0.3$ and $P\{X = 2\} = 0.5$, then $P\{X = 3\} = 0.2$. Thus, $E[X] = (1)(0.3) + (2)(0.5) + (3)(0.2) = 1.9$.

7. The following table shows the possible smaller and larger values with their corresponding probabilities.

Smaller, i	P(i)	Larger, i	P(i)
1	11/36	6	11/36
2	9/36	5	9/36
3	7/36	4	7/36
4	5/36	3	5/36
5	3/36	2	3/36
6	1/36	1	1/36

(a) From the table, the expected value of the smaller of the two upturned faces of the dice is $E[i] = (1)(11/36) + (2)(9/36) + \dots + (6)(1/36) = 91/36 = 2.5278$.

(b) From the table, the expected value of the larger of the two upturned faces of the dice is $E[i] = (6)(11/36) + (5)(9/36) + \ldots + (1)(1/36) = 161/36 = 4.4722$.

9. Let X = gross profit. Then, the expected gross profit is

$E[X] = (\$0)(0.6) + (\$3,000)((0.4)(0.6) + (\$6,000)(0.4)90.4) = \$1,680$.
Thus, the expected net profit = $\$1,680 - \$800 = \$880$.

11. (a) The probability distribution for X is as follows: $P[X = 0] = 38/87$, $P[X = 1] = 40/87$ and $P[X = 2] = 9/87$. Hence $E[X] = (0)(38/87) + (1)(40/87) + (2)(9/87) = 58/87 = 2/3$.

(b) The probability distribution for Y is as follows: $P[Y = 0] = 9/87$, $P[Y = 1] = 40/87$ and $P[Y = 2] = 38/87$. Hence $E[Y] = (0)(9/87) + (1)(40/87) + (2)(38/87) = 4/3$

(c) $E[X + Y] = E[X] + E[Y] = 2/3 + 4/3 = 2$.

13. (a) 1st Location: E[profit] = $(\$4,000)(1/2) - (\$10,000)(1/2) = \$15,000$.

2n location: E[profit] = $(\$60,000)(1/2) - (\$25,000)(1/2) = \$17,500$.

Thus the 2nd location will result in a larger profit.

(b) 1st Location: E[profit] = $(\$40,000)(1/3) - (\$10,000)(2/3)$
$= \$6,666.67$.

2n location: E[profit] = $(\$60,000)(1/3) - (\$25,000)(2/3)$
$= \$3,333.33$.

Thus the 1st location will result in a larger profit.

15. E[gain] = $(\$400)(1/10) - (\$50)(9/10) = -\$5$ (i.e. a loss of $5).

17. (a) E[cost] = $(\$950)(0.1) = \$95 > \$40$ (test cost). Component should not be installed without testing.

(b) E[cost] = $(\$950)(0.05) = \$47.5 > \$40$ (test cost). Component should not be installed without testing.

(c) E[cost] = ($950)(0.01) = $9.5 < $40 (test cost). Component should be installed without testing.

(d) P(defective component) = 40/950 = 4/95 = 0.0421.

19. E[gain] =($99)(1/500) + ($49)(2/500) + ($24)(4/500) – (1)(493/500)

$$= -\$0.40.$$

21. Let F represent player I and S represent player II. The possible ways of winning is represented in S = {FF, FSF, SFF, SS, SFS, FSS}. Thus,

E{number of sets] = (2)(2)(1/2)(1/2) + (3)(4)(1/2)(1/2)(1/2) = 2.5.

23. E[profit] = $1,400 - $125 = $1,275.

25. Since E[X] = μ, then E[X - μ] = E[X] – E[μ] = μ - μ = 0.

27. E[demand] = (0.5)(1,200) + (0.2)(1,500) + (0.3)(1,800) = 1,440 trees.

(a) E[profit] = ($14)(1,200) = $16,800.

(b) E[profit] = [$14×1,200 - $6×300] ×0.5 + [$14×1,500] ×0.2 + [$14×1,500] ×0.3 = $18,000.

(c) E[profit] = [$14×1,200 - $6×600] ×0.5
                + [$14×1,500 - $6×300] ×0.2
                + [$14×1,800] ×0.3 = $18,000.

29. E[demand] = (0)(0.15) + (1)(0.25) + (2)(0.3) + (3)(0.15) + (4)(0.15)
                = 1.9 ≈ 2.

With 2 cakes: E[profit] = [$(0×16 - 2×4)×0.15 + (1×16 - 1×4)×0.25 +
                (2×16 - 0×4)×0.3] = $11.4

With 3 cakes: E[profit] = [$(0×16 - 3×4)×0.15 + (1×16 - 2×4)×0.25 +
                (2×16 - 1×4)×0.3 + (3×16 - 0×4)×0.15]
                = $15.8.

With 4 cakes: E[profit] = [\$(0×16 - 4×4)×0.15 + (1×16 - 3×4)×0.25 +
(2×16 - 2×4)×0.3 + (3×16 - 1×4)×0.15 +
(4×16 - 0×4)×0.15] = \$22.

Thus for highest expected profit they should bake 4 cakes.

31. (a) E[sum] = (2)(1/36) + (3)(2/36) +(4)(3/36) + (5)(4/36) + (6)(5/36) +
(7)(6/36) + (8)(5/36) + (9)(4/36) + (10)(3/36) + (11)(2/36)
+ (12)(1/36) = 7.

(b) Let X = the number on the up side of die I (say) and let Y = the number on the up side of die II (say). From Example 5.5, E[X] = 3.5. Similarly, E[Y] = 3.5. Since X + Y will represent the sum of the numbers on the two up side faces of the dice, then E[sum] = E[X + Y] = E[X] + E[Y] = 3.5 + 3.5 = 7.

33. E[number of failures] = 2×[8×1/8 + 6×1/8 + 1×1/8 + 3×1/8 + 8×1/8 +
7×1/8 + 4×1/8 + 11×1/8]
= 2×6 = 12.

35. Let X = number of traffic tickets per month per taxi. Then,

E[X] = 4×[(0)(0.3) + (1)(0.5) + (2)(0.2)] = 3.6 tickets.

SECTION 5.4 – VARIANCE OF RANDOM VARIABLES

PROBLEMS

1. $Var(U) = 0$; $Var(V) = (-1)^2(1/2) + (1)^2(1/2) - (0) = 1$;

   $Var(W) = (-10)^2(1/2) + (10)^2(1/2) - (0) = 100$;

   Note: In the above, $E[U] = E[V] = E[W] = 0$.

3. $E[X] = c$; $Var[X] = (c)^2(1) - (c)^2 = 0$.

5. $E[X] = 1.9$; $Var[X] = (1)(0.3) + (4)(0.5) + (9)(0.2) - (1.9)^2 = 0.49$.

7. Let X = number of sets played. The possible values of X are 2 or 3 with

   $P(X = 2) = 1/2$ and $P(X = 3) = 1/2$. Thus, $E[X] = 2.5$ and

   $Var(X) = (4)(1/2) + (9)(1/2) - (2.5)^2 = 0.25$.

9. (a) $Var(X)$ because the weights (probabilities) are more spread out for X.

   (b) $E[X] = (0)(0.4) + (1)(0.2) + (2)(0.4) = 1$.

      $Var(X) = (0)(0.4) + (1)(0.2) + (4)(0.4) - 1 = 0.8$.

   (c) $E[Y] = (0)(0.3) + (1)(0.4) + (2)(0.3) = 1$.

      $Var(Y) = (0)(0.3) + (1)(0.4) + (4)(0.3) - 1 = 0.6$.

11. The sample space $S = \{HH, HT, TH, TT\}$. Let X = number of heads.

   Then $P(X = 0) = 1/4$, $P(X = 1) = 1/2$, and $P(X = 2) = 1/4$.

(a) $E[X] = (0)(1/4) + (1)(1/2) + (2)(1/4) = 1$.

   $Var(X) = (0)(1/4) + (1)(1/2) + (4)(1/4) - 1 = 1/2 = 0.5$.

(b) Let Y be the event of a head on the first toss and let Y equals 1 for this event, otherwise Y equals 0. Let Z be the event of a tail on the first toss and let Z equals 1 for this event, otherwise Z equals 0. Thus, $P(Y = 0) = 1/2$, $P(Y = 1) = 1/2$, $P(Z = 0) = 1/2$, $P(Z = 1) = 1/2$, $E[Y] = 1/2$, $E[Z] = 1/2$, $Var(Y) = 1/4$, and $Var(Z) = 1/4$. Now, we can write $X = Y + Z$ and so, $Var(X) = Var(Y + Z) = Var(Y) + Var(Z) = 1/4 + 1/4 = 1/2 = 0.5$ (since Y and Z are independent).

13. Let X = amount of fee.

(a) $Var(X) = 0$ for fixed fee. That is, there is no variability.

(b) $E[X] = (\$8,000)(0.3) = \$2,400$ and $Var(X) = (8000)^2(0.3) - (2400)^2$

   $= 13,440,000$ and so $SD(X) = \sqrt{(Var(X))} = \$3,666.06$.

15. (a) $E[X] = (1)(2/210) + (2)(15/210) + (3)(37/210) + (4)(90/210)$

   $+ (5)(49/210) + (6)(14/210) + (7)(37/210) = 4.0619$.

(b) $Var(X) = (1)(2/210) + (4)(15/210) + (9)(37/210) + (16)(90/210)$

   $+ (25)(49/210) + (36)(14/210) + (49)(37/210) - (4.0619)^2$

   $= 1.1724$. Thus $SD(X) = 1.0828$.

17. $Var(X) = 4$; $Var(3X) = 9Var(X) = (9)(4) = 36$. $SD(3X) = 6$.

19. $Var(X) = Var(Y) = 1$, and X and Y are independent.

(a) $Var(X + Y) = Var(X) + Var(Y) = 1 + 1 = 2$.

(b) $Var(X - Y) = Var(X) + Var(-Y) = Var(X) + Var(-1Y)$

$$= \text{Var}(X) + (-1)^2 \text{Var}(Y) = \text{Var}(X) + \text{Var}(Y) = 1 + 1 = 2.$$

SECTION 5.5 – BINOMIAL RANDOM VARIABLES

PROBLEMS

1. (a) $4! = 24$ (b) $5! = 120$ (c) $7! = 5040$.

3. $10! = 10 \times 9! = 3,628,800$.

5. X is a binomial random variable wit parameters n = 8 and p = 0.4.

   (a) $P\{X = 3\} = 0.2787$.

   (b) $P\{X = 5\} = 0.1239$.

   (c) $P\{X = 7\} = 0.0079$.

7. Let X = number of hockey games that go into overtime. Then X is a binomial random variable with parameters n = 6 and p = 0.1.

   (a) $P\{X \geq 1\} = 1 - P\{X < 1\} = 1 - P\{X = 0\} = 1 - 0.5314 = 0.4686$.

   (b) $P\{X \leq 1\} = P\{X = 0\} + P\{X = 1\}$ 0.5314 + 0.3543 = 0.8857.

9. Let X = number of incorrect digits received.   Thus X is a binomial random variable with parameters n = 5 and p = 0.1.

   (a) 3 or more.

   (b) $P\{X \geq 3\} = P\{X = 3\} + P\{X = 4\} + P\{X = 5\} = 0.0081 + 0.0005 + 0$

            $= 0.0086$.

11. Let X = number of correct guesses.   Thus X is a binomial random variable with parameters n = 8 and p = 0.5.

$$P\{X \geq 7\} = P\{X = 6\} + P\{X = 7\} + P\{X = 8\}$$

$$= 0.1093 + 0.0313 + 0.0039 = 0.1445$$

13. Let X = number of times 6 appears. Thus X is a binomial random variable with parameters n = 4 and p = 1/6.

   (a) $P\{X \geq 1\} = 1 - P\{X < 1\} = 1 - P\{X = 0\} = 1 - 0.4823 = 0.5177$.

   (b) $P\{X = 1\} = 0.3858$.

   (c) $P\{X \geq 2\} = 1 - P\{X \leq 1\} = 1 - [P\{X = 0\} + P\{X = 1\}$

   $$= 1 - [0.4823 + 0.3858] = 0.1319.$$

15. Let X = number of children receiving two sickle cell genes. Then X is a binomial random variable with parameters n = 3 and p = 1/4.

   (a) $P\{X = 0\} = 0.4219$.

   (b) $P\{X = 1\} = 0.4219$

   (c) $P\{X = 2\} = 0.1406$.

   (d) $P\{X = 3\} = 0.0156$.

17. Let X = size of the face value of the die.

   (a) $E[X] = (20)(1/6) = 10/3$.

   (b) $E[5 \text{ or } 6] = 2E[X] = 20/3$.

   (c) $E[\text{even number}] = (3)E[X] = 10$.

   (d) $E[\text{all except } 6] = 5E[X] = 50/3$.

19. Let X = number of bulbs burning for at least 500 hours. Then X is a binomial random variable with parameters n = 8 and p = 0.9.

(a) $P\{X = 8\} = 0.4305$.

(b) $P\{X = 7\} = 0.3826$.

(c) $E[X] = (8)(0.9) = 7.2$.

(d) $Var(X) = (8)(0.9)(0.1) = 0.72$.

21. Let X = number of murder victims killed with a handgun. Then X is a binomial random variable with parameters $n = 4$ and $p = 0.44$.

(a) $P\{X = 4\} = 0.0375$.

(b) $P\{X = 0\} = 0.0983$.

(c) $P\{X \geq 2\} = 1 - P\{X < 1\} = 1 - [P\{X = 0\} + P\{X = 1\}]$

$= 1 - [0.0983 + 0.3091] = 0.5926$.

(d) $E[X] = (4)(0.44) = 1.76$.

(d) $Var(X) = (4)(0.44)(0.56) = 0.9856$ or $SD(X) = 0.9928$.

23. X is a binomial random variable with expected value 4 and variance 2.4. That is, $np = 4$ and $np(1 - p) = 2.4$. Solving gives $n = 10$ and $p = 0.4$.

(a) $P\{X = 0\} = 0.0060$.

(b) $P\{X = 12\} = 0$.

25. (a) $np = 50$ and $\sqrt{[np(1 - p)]} = 5$.

(b) $np = 40$ and $\sqrt{[np(1 - p)]} = 4.8990$.

(c) $np = 60$ and $\sqrt{[np(1 - p)]} = 4.8990$.

(d) $np = 25$ and $\sqrt{[np(1 - p)]} = 3.5355$.

(e) $np = 75$ and $\sqrt{[np(1 - p)]} = 6.1237$.

(f) $np = 50$ and $\sqrt{[np(1 - p)]} = 6.1237$.

SECTION 5.6 – HYPERGEOMETRIC RANDOM VARIABLES

PROBLEMS

1. X is a hypergeometric random variable with parameters $N = 200$, $n = 20$, and $np = 18/200 = 0.09$.

3. X is a hypergeometric random variable with parameters $N = 54$, $n = 6$, and $p = 6/54 = 0.1111$.

4. X is a hypergeometric random variable with parameters $N = 100$, $n = 20$, and $p = 5/100 = 0.05$.

5. X is a binomial random variable with parameters $n = 10$, and $p = 4/52$.

SECTION 5.7 – POISSON RANDOM VARIABLES

PROBLEMS

1. Given that X is a Poisson random variable with mean, $\lambda = 4$.

    (a) $P\{X = 1\} = 0.0733$.

    (b) $P\{X = 2\} = 0.1465$.

    (c) $P\{X > 2\} = 1 - P\{X \leq 2\} = 1 - 0.2381 = 0.7619$.

3. Let X = number of prizes won. Then X is approximately Poisson random variable with mean = $\lambda = (500)(1/1000) = 0.5$.

    (a) $P\{X = 0\} = 0.6065$.

    (b) $P\{X = 1\} = 0.3033$.

    (c) $P\{X \geq 2\} = 1 - P\{X \leq 1\} = 1 - 0.9098 = 0.0902$.

5. Let X = number of claims paid out per month. Then X is a Poisson random variable with mean = $\lambda = 4$.

    (a) $P\{X = 0\} = 0.0183$.

    (b) $P\{X \leq 2\} = 0.2381$.

    (c) $P\{X \geq 4\} = 1 - P\{X \leq 3\} = 1 - 0.4335$.

REVIEW PROBLEMS

1. (a) Since $P\{X \le 4\} = 0.8$ and $P\{X = 4\} = 0.2$, then $P\{X \le 3\} = 0.8 - 0.2$

   $= 0.6$. So $P\{X \ge 4\} = 1 - 0.6 = 0.4$.

   (b) $P\{X < 4\} = P\{X \le 3\} = 0.6$.

3. Let $X =$ number of times he attempts the bar exam.

   (a) The possible values of X are 1, 2, 3, and 4.

   (b) The probability distribution for X is given in the following table.

i	1	2	3	4
$P\{X = i\}$	0.3	0.21	0.147	0.343

   For example $P\{X = 2\} = P\{$failing the first time and passing the second time$\} = (0.7)(0.3) = 0.21$.

   Note: To compute $P\{X = 4\}$ you will have to consider whether he fails on the fourth try or whether he passes.

   (c) $P\{$passing$\} = P\{$passing on first or second or third or fourth attempt$\}$

   $= P\{X = 1\} + P\{X = 2\} + P\{X = 3\} + P\{$passing on the fourth$\}$

   $= 0.3 + 0.21 + 0.147 + (0.7)(0.7)(0.7)(0.3) = 0.7599$.

   (d) $E[X] = (1)(0.3) + (2)(0.21) + (3)(0.147) + (4)(0.343) = 2.533$.

   (e) $Var(X) = (1)(0.3) + (4)(0.21) + (9)(0.147) + (16)(0.343) - (2.533)^2$

   $= 1.5349$.

5. Let X = gambler's final winnings.

    (a) P{X > 0} = P{winning on 1ˢᵗ try or loosing on 1ˢᵗ try and winning on the 2ⁿᵈ try}

    $$= (18/38) + (20/38)(18/38) = 0.7230.$$

    (b) Since P{X > 0} = 0.723, the odds are in your favor. Thus the strategy
         seems to be a "winning strategy" based on this probability.

    (c) E[X] = (1)(18/38) + (2 − 1)(20/38)(18/38) − 3(20/38)(20/38)

    $$= -\$0.108.$$

    That is, the expected winning (loss) is approximately 11 cents. So, based on this expected value, if you play this game for a long period of time, you will end up loosing which is different based on the P{X > 0}.

7. Let X = amount of money the customer spends.

    (a) The probability distribution for X is shown in the table below.

i	0	4,000	6,000
P{X = i}	0.7	0.15	0.15

    (b) E[X] = $[0×0.7 + 4,000×0.15 + 6,000×0.15] = $1,500

    (c) Var(X) = [0 + 4,000²×0.15 + 6,000²×0.15] - 1,500²

    $$= 5,550,000 \text{ (square dollars)}$$

    (d) SD(X) = $2,355.8438.

9. High bid: E[profit] = [25% of 140,000](0.15) = $5,250.

Low bid: E[profit] = [10% of 140,000]0.4) = $5,600.

Thus, the low bid will maximize the expected profit.

11. Let X = total dollar value of all sales.

(a) P{X = 0} = 1/3.

(b) P{X = 500} = P{1 sale and standard} = (1/2)(1/2) = ¼.

(c) P{X = 1000} = P{selling 1 deluxe or 2 standards}

$$= P\{\text{selling 1 deluxe}\} + P\{\text{selling 2 standards}\}$$

$$= (1/2)(1/2) + (1/6)(1/2)(1/2) = 7/24.$$

(d) P{X = 1500} = P{(selling 1 standard and 1 deluxe) or (selling 1 deluxe and 1standard}

$$= 2\times P\{\text{selling 1 standard and 1 deluxe}\}$$

$$= 2\times(1/2)(1/2)(1/6) = (1/12).$$

(e) P{X = 2000} = P{selling 2 deluxe} = (1/2)(1/2)(1/6) = 1/24.

(f) E[X] = $[(0)(1/3) + (500)(1/4) + (1000)(7/24) + (1500)(1/12) +

$$(2000)(1/24)] = \$625.$$

(g) Let Y = amount of money salesman earns, then

E[Y] = $[(625)(0.2)] = $125.

13. (a) E[A] = $[20,000(1/4) + 10,000(1/4) – 15,000(1/2)] = $0.00.

(b) E[B] = $[30,000(1/4) + 20,000(1/4) – 15,000(3/8) – 40,000(1/4)]

$$= -\$6,875 \text{ (loss of } \$6,875).$$

(c) E[A + B] = $0 - $6,875 = -$6,875.

15. $E[X] = \mu$, $SD(X) = \sigma$, and $Y = (X - \mu)/\sigma$

    (a) $E[Y] = E[(X - \mu)/\sigma] = (1/\sigma)E[(X - \mu)] = (1/\sigma)[E(X) - E(\mu)]$

    $= (1/\sigma)[\mu - \mu] = 0$.

    (b) $Var[Y] = Var[(X - \mu)/\sigma] = (1/\sigma^2)Var[(X - \mu)]$

    $= (1/\sigma^2)[Var(X) - Var(\mu)] = = (1/\sigma^2)[\sigma^2 - 0] = 1$.

17. (a) $n = 1$; P{number of heads > number tails} = 0.6.

    (b) $n = 3$; P{number of heads > number tails} = P{2 or 3 heads}

    $= 3(0.6)(0.6)(0.4) + (0.6)(0.6)(0.6) = 0.648$.

    (c) $n = 5$; P{number of heads > number tails}

    $= $ P{3 or 4 or 5 heads} = P{3 heads} + P{4 heads} + P{5 heads}

    $= 0.3456 + 0.2592 + 0.0778 = 0.6826$.

    (d) $n = 7$; P{number of heads > number tails}

    $= $ P{4 or 5 or 6 or 7 heads}

    $= $ P{4 heads} + P{5 heads} + P{6 heads} + P{7 heads}

    $= 0.2903 + 0.2613 + 0.1306 + 0.0280 = 0.7102$.

    (e) $n = 9$; P{number of heads > number of tails}

    $= $ P[5 or 6 or 7 or 8 or 9 heads} = 0.7334.

    (f) $n = 19$; P{number of heads > number of tails}

    $= $ P{10 or 11 or 12 or … or 19 heads} = 0.8139.

19. Let X = number of sales per month. Then, X is a binomial random variable with n = 3 and p = 0.6.

    (a) P{no sales} = PX = 0} = 0.064.

(b) P{2 sales} = P{X = 2} = 0.432.

(c) P{at least one sale} = 1 – P{no sales} = 1 – 0.064 = 0.936, thus

P{at least one sale in the next 3 months} = $(0.936)^3$ = 0.8200.

Chapter 6   NORMAL RANDOM VARIABLES

SECTION 6.2 CONTINUOUS RANDOM VARIABLES

PROBLEMS

1. Let X = number of minutes to complete a repair.

   (a) P{X < 20} = 0.14 + 0.15 = 0.29.

   (b) P{X < 40} = 0.14 + 0.15 + 0.14 + 0.13 = 0.56.

   (c) P{X > 50} = 0.09 + 0.07 + 0.17 = 0.33.

   > Note: P{X > 70} = 0.17 since the total probability under the curve equals 1.

   (d) P{40 < X < 70} == 0.11 + 0.09 + 0.7 = 0.27.

3. Given that the random variable X is uniformly distributed over the interval (0, 1), then

   (a) P{X > 1/3} = 2/3.

   (b) P{X < 0.7} = 0.7

   (c) P{0.3 < X < 0.9} = 0.6.

   (d) P{0.2 <. X < 0.8} = 0.6.

5. (a) P{you are the first to arrive} = P{friend arrive after 2 p.m.}

   = 60/90 = 2/3.

(b) P{ friend will have to wait more than 15 minutes} =

= P{friend arrives between 1:30 p.m. and 1:45 p.m.}

= 15/90 = 1/6.

(c) P{you will have to wait over 30 minutes}

P{friend arrives after 2:30 p.m.}

= 30/90 = 1/3.

7. Let X = number of minutes of playing time for the basketball

player.

(a) $P\{20 < X < 30\} = 0.5$.

(b) $P\{X > 50\} = 0$.

(c) $P\{20 < X < 40\} = 0.75$

(d) $P\{15 < X < 25\} = 0.25/2 + 0.5/2 = 0.375$.

SECTION 6.3 NORMAL RANDOM VARIABLES

PROBLEMS

1. (a) Interval: 128.4 - 19.6 to 128.4 + 19.6 $\Rightarrow$ 108.8 to 148.

    (b) Interval: 128.4 - (2)(19.6) to 128.4 + (2)(19.6) $\Rightarrow$ 89.2 to 167.6.

    (c) Interval: 128.4 - (3)(19.6) to 128.4 + (3)(19.6) $\Rightarrow$ 69.6 to 187.2.

3. (b)

5. (d)

7. (c)

9. (a)

11. (b)

13. (d)

15. (b)

17. (a) $P\{X > 104\} \approx 0.025$ and $P\{Y > 104\} \approx 0.16$. Hence Y is more likely to exceed 104.

    (b) $P\{X > 96\} \approx 0.975$ and $P\{Y > 96\} \approx 0.84$. Hence X is more likely to exceed 96.

    (c) $P\{X > 100\} = 0.5$ and $P\{Y > 100\} = 0.5$. Hence both X and Y will be equally likely to exceed 100.

19. Let X = score on aptitude test.

   (a) P{X > 400} = 0.5, hence do not consider these scores.

   (b) P{X > 450} ≈ 0.34, hence do not consider these scores.

   (c) P{X > 500} ≈ 0.16, hence do not consider these scores.

   (d) P{X > 600} ≈ 0.025, hence consider these scores.

SECTION 6.4   PROBABILITIES ASSOCIATED WITH A STANDARD NORMAL RANDOM VARIABLE

PROBLEMS

1. (a) $P\{Z < 2.2\} = 0.9861$.

(b) $P\{Z > 1.1\} = 1 - P\{Z \le 1.1\} = 1 - 0.8643 = 0.1357$.

(c) $P\{0 < Z < 2\} = P\{Z < 2\} - P\{Z < 0\} = 0.9772 - 0.5 = 0.4772$.

(d) $P\{-0.9 < Z < 1.2\} = P\{Z < 1.2\} - P\{Z < -0.9\}$

   $= P\{Z < 1.2\} - [1 - P\{Z < 0.9\}] = 0.8849 - [1 - 0.8159] = 0.7008$.

(e) $P\{Z > -1.96\} = P\{Z < 1.96\} = 0.975$.

(f) $P\{Z < -0.72\} = 1 - P\{Z < 0.72\} = 1 - 0.7642 = 0.2358$.

(g) $P\{|Z| < 1.64\} = P\{-1.64 < Z < 1.64\}$

   $= P\{Z < 1.64\} - [1 - P\{Z < 1.64\}] = 0.9495 - [1 - 0.9495] = 0.899$.

(h) $P\{|Z| > 1.2\} = P\{Z > 1.2 \text{ or } Z < -1.2\} = P\{Z > 1.2\} + P\{Z < -1.2\}$

   $= 2[1 - P\{Z < 1.2\}] = 2[1 - 0.8849] = 0.2302$.

(i) $P\{-2.2 < Z < 1.2\} = P\{Z < 1.2\} - P\{Z < -2.2\}$

   $= P\{Z < 1.2\} - [1 - P\{Z < 2.2\}] = 0.8849 - [1 - 0.9861] = 0.871$.

3. $P\{-3 < Z < -2\} = P\{2 < Z < 3\}$.

5. $P\{-a < Z < a\} = 1 - [P\{Z < -a\} + P\{Z > a\}]$

   $= 1 - [(1 - P\{Z < a\}) + (1 - P\{Z < a\}] = 1 - [2 - 2P\{Z < a\}]$

   $= 2P\{Z < a\} - 1.$

7. (a) $P\{Z > x\} = 0.05 \Rightarrow P\{Z < x\} = 0.95$, for which $x = 1.645 \approx 1.65$.

   (b) $P\{Z > x\} = 0.025 \Rightarrow P\{Z < x\} = 0.975$, for which $x = 1.96$.

   (c) $P\{Z > x\} = 0.005 \Rightarrow P\{Z < x\} = 0.995$, for which $x = 2.575 \approx 2.58$.

   (d) $P\{Z < x\} = 0.5$, for which $x = 0$.

   (e) $P\{Z < x\} = 0.66$, for which $x = 0.41$.

   (f) $P\{|Z| < x\} = 0.99 \Rightarrow P\{-x < Z < x\} = 0.99 \Rightarrow 2P\{Z < x\} - 1 = 0.99.$

   That is, $P\{Z < x\} = 0.995$, for which $x = 2.575 \approx 2.58$.

   (g) $P\{|Z| < x\} = 0.75 \Rightarrow P\{-x < Z < x\} = 0.75 \Rightarrow 2P\{Z < x\} - 1 = 0.75.$

   That is, $P\{Z < x\} = 0.875$, for which $x = 1.15$.

   (h) $P\{|Z| > x\} = 0.9 \Rightarrow P\{Z > x\} + P\{Z < -x\} = 0.9 \Rightarrow (1 - P\{Z < x\})$

   $+ (1 - P\{Z < x\}) = 2 - 2P\{Z < x\} = 0.9.$ That is, $P\{Z < x\} = 0.55$, for

   which $x = 0.12$.

   (i) $P\{|Z| > x\} = 0.5 \Rightarrow P\{Z > x\} + P\{Z < -x\} = 0.5 \Rightarrow (1 - P\{Z < x\}) +$

   $(1 - P\{Z < x\}) = 2 - 2P\{Z < x\} = 0.5.$ That is, $P\{Z < x\} = 0.75$, for

   which $x = 0.67$.

SECTION 6.6 ADDITIVE PROPERTY OF NORMAL RANDOM VARIABLES

PROBLEMS

1. Since x > a, then x - μ > a - μ. Since σ > 0, then (x - μ)/σ > (a - μ)/σ.

3. Let X = length of time (in months) the hair dryer functions before breaking down.

    We need to find P{X < 36}. Now, P{X < 36} = P{Z < (36 - 40)/8} = P{Z < -0.5} = 1 - P{Z < 0.5} = 1 - 0.6915 = 0.3085.

5. Let X = number of bottles sold.

    (a) P{X > 200} = P{Z > (200 - 212)/40} = P{Z > -0.3} = 0.6179.

    (b) P{X < 250} = P{Z < (250 - 212)/40} = P{Z < 0.95} = 0.8289.

    (c) P{200 < X < 250} = P{(200 - 212)/40 < Z < (250 - 212)/40} =
         P{-0.3 < Z < 0.95} = P{Z < 0.95} - P{Z < -0.3} =
         P{Z < 0.95} - [1 - P{Z < 0.3}] = 0.8289 - [1 - 0.6179] = 0.4468.

7. Let X = life of the tire.

    P(X > 50 | X > 40} = P{X > 50 ∩ X > 40}/P{X > 40}

    = P{X > 50}/P{X > 40} = 0.0013/0.1587 = 0.0082.

9. Let X = time (minutes) required to complete loan application form.

    (a) P{X < 75} = P{Z < (75 - 90)/15} = P{Z < -1} = 1 - P{Z < 1}

         = 1 - 0.8413 = 0.1587.

(b) $P\{X > 100\} = P\{Z > (100 - 90)/15\} = P\{Z > 0.67\} = 1 - P\{Z < 0.67\}$

   $= 1 - 0.7486 = 0.2514.$

(c) $P\{90 < X < 120\} = P\{(90 - 90)/15 < Z < (120 - 90)/15\}$

   $= P\{0 < Z < 2\} = P\{Z < 2\} - P\{Z < 0\} = 0.9772 - 0.5 = 0.4772.$

11. Let X = activation pressure of the valve.

   $P\{20 < X < 32\} = P\{(20 - 26)/4 < Z < (32 - 26)/4\} = P\{-1.5 < Z < 1.5\}$

   $= P\{Z < 1.5\} - P\{Z < -1.5\} = P\{Z < 1.5\} - [1 - P\{Z < 1.5\}]$

   $= 0.9332 - [1 - 0.9332] = 0.8664.$

13. Let X = lifetime (years) of a color TV picture tube.

   (a) $P\{X > 10\} = P\{Z > (10 - 8.2)/1.4\} = P\{Z > 1.29\} = 1 - P\{Z < 1.29\}$

   $= 1 - 0.9015 = 0.0985.$

   (b) $P\{X < 5\} = P\{Z < (5 - 8.2)/1.4\} = P\{Z < -2.29\} = 1 - P\{Z < 2.29\}$

   $= 1 - 0.9890 = 0.011.$

   (c) $P\{5 < X < 10\} = P\{-2.29 < Z < 1.29\} = P\{Z < 1.29\} - P\{Z < -2.29\}$

   $= 0.9015 - 0.011 = 0.8905$

15. Let X = height of an adult woman in the US.

(a) $P\{X < 63\} = P\{Z < (63 - 64.5)72.4\} = P\{Z < -0.63\}$

$- 1 - P\{Z < 0.63\} = 1 - 0.7357 = 0.2643.$

(b) $P\{X < 70\} = P\{Z < (70 - 64.5)72.4\} = P\{Z < 2.29\} = 0.9890.$

(c) $P\{63 < X < 70\} = P\{X < 70\} - P\{X < 63\} = 0.9890 - 0.2643$

$= 0.7247.$

(d) $P\{X < 72\} = P\{Z < (72 - 64.5)/2.4\} = P\{Z < 3.13\} = 0.9991.$ Thus,

99.91% of the women are shorter than Alice.

(e) Let Y = height of a second female. Thus, we need

$P\{(X + Y)/2 > 67.5\}$ or $P\{(X + Y) > 135\}$. Now, the mean for

$(X + Y)$ is $(64.5)(2) = 129$ and the standard deviation for $(X + Y)$ is

$\sqrt{2.4^2 + 2.4^2} = 3.39.$ So, $P\{(X + Y) > 135\}$

$= P\{Z > (135 - 129)/3.39\} = P\{Z > 1.77\} = 1 - P\{Z < 1.77\}$

$= 1 - 0.9616 = 0.0384.$

SECTION 6.7   PERCENTILES OF NORMAL RANDOM VARIABLES

PROBLEMS

1. (a) $z_{0.07} = 1.48$.   (b) $z_{0.12} = 1.18$.   (c) $z_{0.3} = 0.52$.   (d) $z_{0.03} = 1.88$.

   (e) $z_{0.65} = -0.39$.   (f) $z_{0.05} = 0$.   (g) $z_{0.95} = -1.65$.   (h) $z_{0.008} = 2.41$.

3. (a) $P\{X > x\} = 0.5 \Rightarrow P\{Z > (x - 50)/6\} = 0.5 \Rightarrow (x - 50)/6 = 0$ or $x = 50$.

   (b) $P\{X > x\} = 0.1 \Rightarrow P\{Z > (x - 50)/6\} = 0.1 \Rightarrow (x - 50)/6 = 1.28$ or

   $x = 57.68$.

   (c) $P\{X > x\} = 0.025 \Rightarrow P\{Z > (x - 50)/6\} = 0.025 \Rightarrow (x - 50)/6 = 1.96$

   or $x = 61.76$.

   (d) $P\{X < x\} = 0.05 \Rightarrow P\{Z < (x - 50)/6\} = 0.05 \Rightarrow (x - 50)/6 = -1.645$

   or $x = 40.13$.

   (e) $P\{X < x\} = 0.88 \Rightarrow P\{Z > (x - 50)/6\} = 0.88 \Rightarrow (x - 50)/6 = 1.175$ or
   $x = 57.05$.

5. Let X = real estate brokers exam scores and let x be the required cutoff
   score. Thus, we have $P\{X > x\} = 0.25$ or $P\{Z > (x - 420)/66)\} = 0.25$.
   That is, $(x - 420)/66 = 0.67$ or $x = 464.22$.

   The passing scores should begin at 465.

7. Let X = time (seconds) it takes to run a mile by the high school boys and
   let x be the cutoff time (seconds). Thus, we have $P\{X < x\} = 0.05$. That
   is, $P\{Z < (x - 460)/40\} = 0.05$ or $(x - 460)/40 = -1.645$ from which
   $x = 394.2$ seconds.

9. Let X = amount of radiation that can be absorbed before death ensues.

Let x be the amount of radiation above which 5% will survive. Thus, $P\{X > x\} = 0.05$. That is, $P\{X > x\} = 0.05 \Rightarrow P\{Z > (x - 500)/150\} = 0.05$ from which $(x - 500)/150 = 1.645$. Solving, gives $x = 746.75$ (roentgens).

11. Let X = amount of people attending the home football games .

   (a) $P\{X > 46,000\} = P\{Z > (46,000 - 52,000/4,000\} = P\{Z > -1.5\} =$

   $P\{Z < 1.5\} = 0.9332$. Thus, 93.32% of the games have over

   46,000. Hence the statement is true.

   (b) $P\{X\ 58,000\} = P\{Z > (58,000 - 52,000)/4,000\} = P\{Z > 1.5\} =$
   $1 - P\{Z < 1.5\} = 1 - 0.9332 = 0.0668$. Thus 6.68% of the games
   have over 58,000. Thus the statement is true.

13. Let X = fasting blood glucose level of diabetics (100 ml of blood).

Let x be the value of the fasting blood glucose level for a diabetic to be in the lower 20%. Thus we have $P\{X < x\} = 0.2$ or $P\{Z < (x - 106)/8\} = 0.2$. That is, $(x - 106)/8 = -0.84$ from which $x = 99.28$.

REVIEW PROBLEMS

1. Let X = heights of adult males.

    (a) P{X > 65} = P{Z > (65 - 69)/2.8} = P{Z > -1.43} = P{Z < 1.43}

    = 0.9236.

    (b) P{62 < X < 72} = P{(62 - 69)/2.8 < Z < (72 - 69)/2.8} =

    P{-2.5 < Z < 1.07} = P{Z < 1.07} - P{Z < -2.5} =

    P{Z < 1.07} - [1 - P{Z < 2.5}] = 0.8577 - [1 - 0.9938] = 0.8515.

    (c) P{|X - 69| > 6} = P{|Z| > 6/2.8} = P{|Z| > 2.14} = 2P{Z > 2.14} =

    2[1 - P{Z < 2.14}] = 2[1 - 0.9838] = 0.0324.

    (d) P{63 < X < 75} = P{(63 - 69)/2.8 < Z < (75 - 69)/2.8} =

    P{-2.14 < Z < 2.14} = P{Z < 2.14} - P{Z < -2.14} =

    P{Z < 2.14} - [1 - P{Z < 2.14}] = 0.9838 - [1 - 0.9838] = 0.9676.

    (e) P{X > 72} = P{Z > (72 - 69)/2.8} = P{Z > 1.07} = 1 - P{Z < 1.07} =

    1 - 0.8577 = 0.1423.

    (f) P{X < 60} = P{Z < (60 - 69)/2.8} = P{Z < -3.21} = 1 - P{Z < 3.21}

    = 1 - 0.9993 = 0.0007.

(g) $P\{X > x\} = 0.01 \Rightarrow P\{Z > (x - 69)/2.8\} = 0.01$ or $(x - 69)/2.8 = 2.33$
from which x = 75.524.

(h) $P\{X < x\} = 0.95 \Rightarrow P\{Z < (x - 69)/2.8\} = 0.95$ or $(x - 69)/2.8 = 1.645$
from which x = 73.606.

(i) $P\{X < x\} = 0.40 \Rightarrow P\{Z < (x - 69)/2.8\} = 0.4$ or $(x - 69)/2.8 = -0.25$
from which x = 68.3.

3. Let X = jet pilots' blackout thresholds and x be the cutoff threshold.
Thus we have $P\{X > x\} = 0.25$. That is, $P\{Z > (x - 4.5)/0.7\} = 0.25$ or
$(x - 4.5)/0.7 = 0.67$ from which x = 4.969.

5. Let X = life (hours) of the light bulb.

(a) $P\{X > 560\} = P\{Z > (560 - 500)/60\} = P\{Z > 1\} = 1 - P\{Z < 1\} =$
$1 - 0.8413 = 0.1587$. Thus 15.87% of the bulbs will last more than
560 hours.

(b) $P\{X < 440\} = P\{Z < (440 - 500)/60\} = P\{Z < -1\} = 1 - P\{Z < 1\} =$
$1 - 0.8413 = 0.1587$. Thus 15.87% of the light bulbs will last less
than 440 hours.

(c) $P\{X > 560 \mid X > 440\} = P\{(X > 560) \cap (X > 440)\}/P\{X > 440\} =$

$P\{X > 560\}/P\{X > 440\} = 0.1587/(1 - 0.1587) == 0.1886$.

(d) Let x be the cutoff value, then we have $P\{X > x\} = 0.1$ or
$P\{Z > (x - 500)/60\} = 0.1$. Thus $(x - 500)/60 = 1.28$ or x == 576.8
(hours).

7. Let X = yearly cost of upkeep for the 1st year and Y = yearly cost of
upkeep for the 2nd year. The random variable here is X + Y which is
normal with a mean of $6,000 and a standard deviation of $848.5281.

(a) $P\{(X+Y)>5{,}000\} = P\{Z>(5{,}000-6{,}000)/848.5281\} =$

$P\{Z>-1.18\} = P\{Z<1.18\} = 0.8810.$

(b) $P\{(X+Y)<7{,}000\} = P\{Z<(7{,}000-6{,}000)/848.5281\} =$

$P\{Z<1.18\} = 0.8810.$

(c) $P\{5{,}000<(X+Y)<7{,}000\} = P\{-1.18<Z<1.18\} =$

$P\{Z<1.18\} - P\{Z<-1.18\} = 0.8810 - [1-0.8810] = 0.762.$

9. Let X = gross weekly sales in the 1st week and Y = gross weekly sales in the 2nd week. Thus $(X+Y)$ will be normal with a mean \$37,600 and a standard deviation of 12,727.9206 when X and Y are independent.

(a) $P\{X>20{,}000\} = P\{Z>(20{,}000-18{,}800)/9{,}000\} = P\{Z>0.13\} =$

$1 - P\{Z<0.13\} = 1 - 0.5517 = 0.4483.$

(b) $P\{$ sales will exceed \$20,000 in two successive weeks$\} =$

$P\{(X>20{,}000) \cap (Y>20{,}000)\} = [P\{X>20{,}000\}]\times[P\{Y>20{,}000\}]$
$= (0.4483)(0.4483) = 0.2010.$

(c) $P\{(X+Y)>40{,}000\} = P\{Z>(40{,}000-37{,}600)/12{,}727.9206\} =$

$P\{Z>0.19\} = 1 - P\{Z<0.19\} = 1 - 0.5753 = 0.4247.$

11. Let X = difference in the number of points scored between the team that is favored and the opposing team.

(a) $P\{X>0\} = P\{Z>(0-7)/14\} = P\{Z>-0.5\} = P\{Z<0.5\} = 0.6915.$

(b) $P\{X<0\} = P\{Z<(0-4)/14\} = P\{Z<-0.29\} = 1 - P\{Z<0.29\} =$

$$1 - 0.6141 = 0.3859.$$

(c) $P\{X < 0\} = P\{Z < (0 - 14)/14\} = P\{Z < -1\} = 1 - P\{Z < 1\} =$

$$1 - 0.8413 = 0.1587.$$

13. Let X = amount (pounds) of tomatoes consumed per year by a randomly chosen person.

Let F be the event of a female and M be the event of a male.

(a) $P\{(X < 14) \cap F\} = P\{(X < 14) \mid F\} \times P\{F\}$. Now $P\{(X < 14) \mid F\} =$

$P\{Z < (14 - 14)/2.7\} = P\{Z < 0\} = 0.5$ and $P\{F\} = 0.5$, hence

$$P\{(X < 14) \cap F\} = (0.5)(0.5) = 0.25.$$

(b) $P\{(X > 14) \cap M\} == P\{(X > 14) \mid M\} \times P\{M\}$. Now $P\{(X > 14) \mid M\}$

$P\{Z > (14 - 14.6)/3\} = P\{Z > -0.2\} = P\{Z < 0.2\} = 0.5793$ and

$P\{M\} = 0.5$, hence $P\{(X > 14) \cap M\} = (0.5793)(0.5) = 0.2897.$

Chapter 7   DISTRIBUTIONS OF SAMPLING STATISTICS

SECTION 7.3 SAMPLE MEAN

PROBLEMS

1. (a) Probability distribution of the means when n = 3.

$\overline{X}$	1	4/3 ($\approx$1.333)	5/3 ($\approx$1.6667)	2
P{$\overline{X}$}	0.125	0.375	0.375	0.125

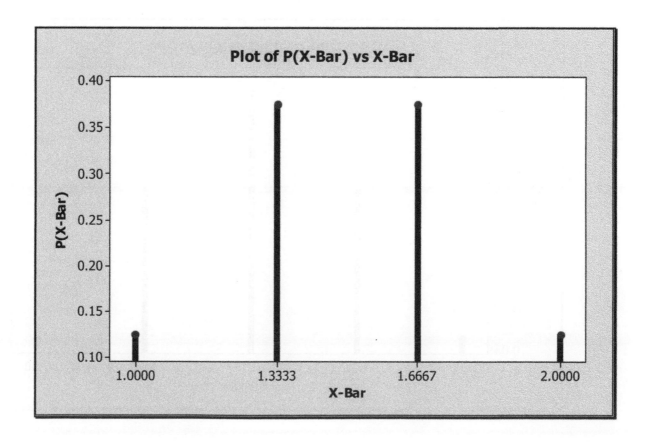

$$\mu = E[\overline{X}] = (1)(0.125) + (4/3)(0.375) + (5/3)(0.375) + (2)(0.125) = 1.5.$$

$$Var(\overline{X}) = E[(\overline{X} - \mu)^2] = (1 - 1.5)^2(0.125) + (4/3 - 1.5)^2(0.375)$$

$$+ (5/3 - 1.5)^2(0.375) + (2 - 1.5)^2(0.125)$$

$$= 0.0833. \text{ Hence the } SD(\overline{X}) = \sqrt{(0.0833)} = 0.2886.$$

(b) Probability distribution for the means when n = 4.

$\overline{X}$	1	1.25	1.5	1.75	2
$P\{\overline{X}\}$	0.0625	0.25	0.375	0.25	0.0625

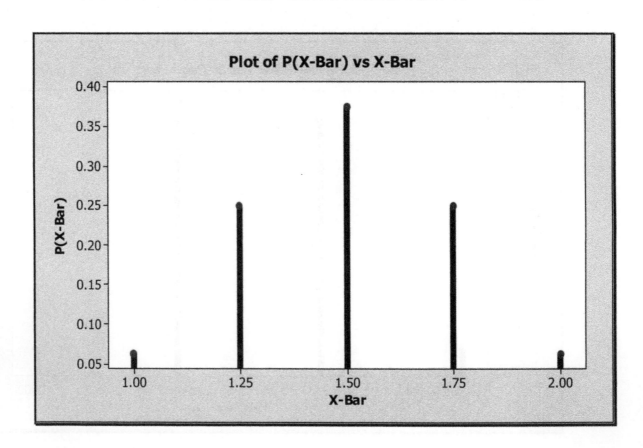

$$\mu = E[\overline{X}] = (1)(0.0625) + (1.25)(0.25) + (1.5)(0.375) + (1.75)(0.25)$$
$$+(2)(0.0625)=1.5.$$

$$Var(\overline{X}) = E[(\overline{X} - \mu)^2] = (1 - 1.5)^2(0.0625) + (1.25 - 1.5)^2(0.25)$$

$$+ (1.5 - 1.5)^2(0.375) + (1.75 - 1.5)^2(0.25)$$

$$+ (2-1.5)^2(0.0625) = 0.0625.$$

Hence the $SD(X) = \sqrt{(0.0625)} = 0.25$.

3. (a) $\mu = E[X] = (1)(1/3) + (2)(1/3) + (3)(1/3) = 2$.

   (b) $\sigma^2 = Var(X) = E[(X - \mu)^2] = (1 - 2)^2(1/3) + (2 - 2)^2(1/3) + (3 - 2)^2(1/3)$

   $$= 2/3.$$

   Hence the $SD(X) = \sqrt{(2/3)} = 0.8165$.

   (c) The possible outcomes are the pairs (1, 1), (1, 2), (1, 3), (2, 2), (2, 1),

   (2, 3), (3, 3), (3, 2) and (3, 1). Thus the possible values of X are 1,

   1.5, 2, 2.5 and 3. Now,

   (d) $P\{\overline{X} = 1\} = P\{(1, 1)\} = 1/9$.

   $P\{\overline{X} = 1.5\} = P\{(1, 2) \text{ or } (2, 1)\} = 2/9$.

   $P\{\overline{X} = 2\} = P\{(1, 3) \text{ or } (2, 2) \text{ or } (3, 1)\} = 3/9$.

   $P\{\overline{X} = 2.5\} = P\{(2, 3) \text{ or } (3, 2)\} = 2/9$.

   $P\{\overline{X} = 3\} = P\{(3, 3)\} = 1/9$.

(e) $E[\overline{X}] = (1)(1/9) + (1.5)(2/9) + (2)(3/9) + (2.5)(2/9) + (3)(1/9) = 2.$

$Var(\overline{X}) = E[(\overline{X} - \mu)^2] = (1 - 2)^2(1/9) + (1.5 - 2)^2(2/9) + ...$

$$+ (3 - 2)^2(1/9) = 1/3.$$

$SD(\overline{X}) = \sqrt{(1/3)} = 0.5774.$

Yes. The answers are consistent. (i) $E[\overline{X}] = 2 = \mu.$

(ii) $Var(\overline{X}) = 1/3 = \sigma^2/n = (2/3)/2.$

5. (a) $E[\overline{X}] = 2.4mg;$ $Var(\overline{X}) = (0.2)^2/36 = 0.0011.$

(b) $E[\overline{X}] = 2.4$ mg; $Var(\overline{X}) = (0.2)^2/64 = 0.000625.$

(c) $E[\overline{X}] = 2.4$ mg; $Var(\overline{X}) = (0.2)^2/100 = 0.0004.$

(d) $E[\overline{X}] = 2.4$ mg; $Var(\overline{X}) = (0.2)^2/900 = 0.000044.$

7. $E[\overline{X}] = 155$ pounds; $SD(\overline{X}) = 28/\sqrt{(100)} = 2.8$ pounds.

The average of the total weight of the passenger

$= (155)(100) = 15,500$ pounds.

The standard deviation of the total weight $= (100)(2.8) = 280$ pounds.

SECTION 7.4   CENTRAL LIMIT THEOREM

PROBLEMS

1. (a) $P\{124 \leq \overline{X} \leq 132\} = P\{(124 - 128)/(16/3) \leq Z \leq (132 - 128)/(16/3)\}$

    $= P\{-0.75 \leq Z \leq 0.75\} = 2P\{Z \leq 0.75\} - 1 = 2(0.7734) - 1 = 0.5468.$

    (b) $P\{124 \leq \overline{X} \leq 132\} = P\{(124 - 128)/(16/5) \leq Z \leq (132 - 128)/(16/5)\}$

    $= P\{-1.25 \leq Z \leq 1.25\} = 2P\{Z \leq 1.25\} - 1 = 2(0.8944) - 1 = 0.7888.$

    (c) $P\{124 \leq \overline{X} \leq 132\} = P\{(124 - 128)/(16/10) \leq Z \leq (132 - 128)/(16/10)\}$

    $= P\{-2.5 \leq Z \leq 2.5\} = 2P\{Z \leq 2.5\} - 1 = 2(0.9938) - 1 = 0.9876.$

3. Let X = total yearly payout of the insurance company.

    $P\{2500000 \leq X \leq 2700000\} = P\{2500000/10000 \leq \overline{X} \leq 2700000/10000\}$

    $= P\{250 \leq \overline{X} \leq 270\} = P\{(250 - 260)/(800/\sqrt{10000}) \leq Z \leq$

    $(270 - 260)/(800/\sqrt{10,000})\} = P\{-1.25 \leq Z \leq 1.25\} = 2P\{Z < 1.25\} - 1$

    $= 2(0.8944) - 1 = 0.7888.$

5. Let X = amount of time to develop a photographic print.

    (a) $P\{X > 1720\} = P\{\overline{X} > 1720/100\} = P\{\overline{X} > 17.2\}$

    $= P\{Z > (17.2 - 17)/(0.8/\sqrt{100})\} = P\{Z > 2.5\}$

    $= 1 - P\{Z \leq 2.5\} = 1 - 0.9938 = 0.0062.$

(b) $P\{1690 < X < 1710\} = P\{16.9 < \overline{X} < 17.1\} = P\{-1.25 < Z < 1.25\}$

$= P\{Z < 1.25\} - P\{Z < -1.25\} = 2P\{Z < 1.25\} - 1 = 2(0.8944) - 1$

$= 0.7888.$

7. Let X = number of sheets of paper.

$P\{X < 2300\} = P\{\overline{X} < 2300/10\} = P\{\overline{X} < 230\} =$

$P\{Z < (230 - 200)/(50/\sqrt{10})\} = P\{Z < 1.9\} = 0.9713.$

9. Let X = sum of the errors.

$P\{X > 3\} = P\{\overline{X} > 3/50\} = P\{Z > (3/50 - 0)/((1/12)/\sqrt{50})\} =$

$P\{Z > 5.09\} = 1 - P\{Z \le 5.09\} = 1 - 1 = 0.$

11. (a) $P\{-0.5 < \overline{X} - d < 0.5\} = P\{-0.5/(3/\sqrt{100}) < Z < 0.5/(3/\sqrt{100})\}$

$= P\{-1.67 < Z < 1.67\} = 2P\{Z < 1.67\} - 1 = 2(0.9525) - 1 = 0.905.$

(b) $P\{-0.5 < \overline{X} - d < 0.5\} = P\{-0.5/(2/\sqrt{10}) < Z < 0.5/(2/\sqrt{10})\}$

$= P\{-0.79 < Z < 0.79\} = 2P\{Z < 0.79\} - 1 = 2(0.7852) - 1 = 0.5704.$

13. Let X = amount of nicotine per cigarette.

(a) $P\{\overline{X} > 2.5\} = P\{Z > (2.5 - 2.4)/(0.2/\sqrt{100})\} = P\{Z > 5\} = 0.$

(b) $P\{Z < 2.25\} = P\{Z < (2.25 - 2.4)/(0.2/\sqrt{100})\} = P\{Z < -7.5\} = 0.$

15. (a) $P\{96 < \overline{X} < 104\} =$

$P\{(96 - 100)/(16/\sqrt{16}) < Z < (104 - 100)/(16/\sqrt{16})\} =$

$$P\{-1 < Z < 1\} = 2P\{Z < 1\} - 1 = 2(0.8413) - 1 = 0.6826.$$

(b) $P\{96 < \overline{X} < 104\} =$

$$P\{(96 - 100)/(8/\sqrt{16}) < Z < (104 - 100)/(8/\sqrt{16})\} =$$

$$P\{-2 < Z < 2\} = 2P\{Z < 2\} - 1 = 2(0.9772) - 1 = 0.9544.$$

(c) $P\{-4 < Z < 4\} = 1.$

(d) $P\{-8 < Z < 8\} = 1.$

(e) $P\{-16 < Z < 16\} = 1.$

SECTION 7.5   SAMPLING PROPORTIONS FROM A FINITE POPULATION

PROBLEMS

1. (a) $E[\overline{X}] = p = 0.6$; $SD(\overline{X}) = \sqrt{[p(1-p)/n]} = \sqrt{[(0.6)(0.4)/10]} = 0.1549$.

   (b) $E[\overline{X}] = p = 0.6$; $SD(\overline{X}) = \sqrt{[p(1-p)/n]} = \sqrt{[(0.6)(0.4)/100]} = 0.0490$.

   (c) $E[\overline{X}] = p = 0.6$; $SD(\overline{X}) = \sqrt{[p(1-p)/n]} = \sqrt{[(0.6)(0.4)/1000]} = 0.0155$.

   (d) $E[\overline{X}] = p = 0.6$;

   $SD(\overline{X}) = \sqrt{[p(1-p)/n]} = \sqrt{[(0.6)(0.4)/10000]} = 0.0049$.

3. $n = 50$, $p = 0.1$, $np = 5$, $n(1-p) = 45$ so we can use the normal approximation. Let X = number of defectives.

   (a) $P\{X = 0\} \approx P\{-0.5 \le X \le 0.5\} =$

   $P\{(-0.5 - 5)/\sqrt{((50)(0.1)(0.9))} \le Z \le (0.5 - 5)/\sqrt{((50)(0.1)(0.9))}\} =$

   $P\{-2.59 \le Z \le -2.12\} = P\{Z \le 2.59\} - P\{Z \le 2.12\} = 0.9952 - 0.9830$

   $= 0.0122$.

   (b) 15% of 50 = 7.5. Thus $P\{X > 7.5\} \approx P\{X \ge 7\} =$

   $P\{Z > (6.5 - 5)/\sqrt{((50)(0.1)(0.9))}\} = P\{Z > 0.71\} = 1 - P\{Z \le 0.71\}$

   $= 1 - 0.7611 = 0.2389$.

   (c) 8% of 50 = 4, 12% of 50 = 6, so $P\{4 \le X \le 6\} \approx P\{3.5 \le X \le 6.5\} =$

   $P\{-0.71 \le Z \le 0.71\} = 2P\{Z \le 0.71\} - 1 = 2(0.7611) - 1 = 0.5222$.

5. Let X = number of unemployed German workers.

    (a) $P\{X \le 40\} \approx P\{Z < (40.5 - 36.8)/\sqrt{((400)(0.092)(0.908))}\} =$

        $= P\{Z < 0.64\}$

        $= 0.7389.$

    (b) $P\{X > 50\} = P\{X \ge 51\} \approx P\{Z > (50.5 - 36.8)/\sqrt{((400)(0.092)(0.908))}\}$

        $= P\{Z > 2.37\} = 1 - P\{Z \le 2.37\} = 1 - 0.9911 = 0.0089.$

7. Let X = number of unemployed Canadian workers.

    (a) $P\{X \le 10\} \approx P\{Z < (10.5 - 13.8)/\sqrt{((200)(0.069)(0.931))}\} =$

        $P\{Z < -0.9207\} = 1 - P\{Z < 0.9207\} = 1 - 0.8212 = 0.1788.$

    (b) $P\{X > 25\} = P\{X \ge 26\} \approx P\{Z \ge (25.5 - 13.8)/\sqrt{((200)(0.069)(0.931))}\}$

        $= P\{Z \ge 3.2642\} = 1 - P\{Z < 3.2642\} = 1 - 0.9994$

    $= 0.0006.$

9. Let X = number of students who will attend.

    (a) $P\{X > 160\} = P\{X \ge 161\}$

        $\approx P\{Z \ge (160.5 - 140)/\sqrt{((350)(0.4)(0.6))}\} = P\{Z \ge 2.24\}$

        $= 1 - P\{Z < 2.24\} = 1 - 0.9875 = 0.0125.$

    (b) $P\{X < 150\} = P\{X \le 149\}$

        $\approx P\{Z \le (149.5 - 140)/\wedge((350)(0.4)(0.6))\} = P\{Z \wedge 1.04\}$

        $= 0.8508.$

11. Let X = number of students planning to major in arts and humanities.

$$P\{X \geq 22\} \approx P\{Z > (21.5 - 18)/\sqrt{((200)(0.09)(0.91))}\} = P\{Z \geq 0.86\}$$

$$= 1 - P\{Z < 0.86\} = 1 - 0.8051 = 0.1949.$$

13. Let X = number of students planning to major in one of the sciences.

$$P\{X \geq 30\} \approx P\{Z \geq (29.5 - 30)/\sqrt{((200)(0.15)(0.85))}\} = P\{Z \geq -0.1\}$$

$$= 1 - P\{Z < 0.1\} = 1 - 0.5398 = 0.4602.$$

15. (a) $P\{X \leq 25\} = P\{X \leq 25.5\} \approx P\{Z \leq (25.5 - 20)/4\} = P\{Z \leq 1.38\}$

$$= 0.9162.$$

(b) $P\{X > 30\} = P\{X \geq 31\} \approx P\{Z \geq (30.5 - 20)/4\} = P\{Z \geq 2.63\}$

$$= 1 - P\{Z < 2.63\} = 1 - 0.9957 = 0.0043.$$

(c) $P\{15 < X < 22\} = P\{16 \leq X \leq 21\}$

$$\approx P\{(15.5 - 20)/4 \leq Z \leq (21.5 - 20)/4\} = P\{-1.13 \leq Z \leq 0.38\}$$

$$= P\{Z \leq 0.38\} - [1 - P\{Z \leq 1.13\}] = 0.5188.$$

17. Let X = number of students graduating in 4 years.

(a) $P\{X < 250\} = P\{X \leq 249\} \approx P\{Z \leq (249.5 - 230)/11.1445\}$

$$= P\{Z \leq 1.75\} = 0.9599.$$

(b) $P\{175 < X < 225\} = P\{176 \leq X \leq 224\}$

$$\approx P\{(175.5 - 230)/11.1445 \leq Z \leq (224.5 - 230)/11.1445\}$$

$= P\{-4.89 \le Z \le -0.4935\} = P\{Z \le 4.89\} - P\{Z \le 0.49\} = 0.3121.$

19. Let X = number of females who are overweight by 30% or more.

(a) $P\{X \ge 25\} = P\{X \ge 24.5\} \approx P\{Z \ge (24.5 - 41.1)/5.96)$

$= P\{Z \ge -2.79\} = P\{Z \le 2.79\} = 0.9974.$

(b) Let X = number of females who sleep 6 hours or less per night.

$P\{X < 50\} = P\{X \le 49\} \approx P\{Z \le (49.5 - 64.2)/7.1036\}$

$= P\{Z \le -2.07\} = 1 - P\{Z \le 2.1\} = 0.0192.$

SECTION 7.6 DISTRIBUTION OF THE SAMPLE VARIANCE
OF A NORMAL POPULATION

PROBLEMS

1. (a) Chi-square statistic (χ^2) = 5.7 with 4 degrees of freedom.

   (b) Chi-square statistic (χ^2) = 1.181 with 5 degrees of freedom.

   (c) Chi-square statistic (χ^2) = 1.13 with 2 degrees of freedom.

REVIEW PROBLEMS

1. Let X = score of a student on the last Scholastic Aptitude Test.

    (a) $P\{\overline{X} > 507\} = P\{Z > (507 - 517)/(120/12)\} = P\{Z > -1\}$

      $= P\{Z \le 1\} = 0.8413$.

    (b) $P\{\overline{X} > 517\} = P\{Z > (517 - 517)/(120/12)\} = P\{Z > 0\} = 0.5$.

    (c) $P\{\overline{X} > 537\} = P\{Z > (537 - 517)/(120/12)\} = P\{Z > 2\}$

      $= 1 - P\{Z \le 2\} = 1 - 0.9772 = 0.0228$.

    (d) $P\{\overline{X} > 550\} = P\{Z > (550 - 517)/(120/12)\} = P\{Z > 3.3\}$

      $= 1 - P\{Z \le 3.3\} = 1 - 0.9995 = 0.0005$.

3. (a) $\mu = \sum_{i=1}^{4} i \times P\{X = i\} = (1)(0.1)+(2)(0.2)+(3)(0.3)+(4)(0.4) = 3$.

    (b) $\sigma^2 = \text{Var}(X) = \sum_{i=1}^{4} i^2 \times P\{X = i\} - \mu^2$

      $= (1)^2(0.1) + (2)^2(0.2) + (3)^2(0.3) + (4)^2(0.4) - (3)^2 = 1$.

    Thus, $\sigma = \text{SD}(X) = 1$.

    (c) There are sixteen possible samples (with replacement) of size 2 from

      the set 1, 2, 3, and 4. The samples are (1, 1), (1,2), (2,1), (1, 3),

      (3, 1), (1, 4), (4, 1), (2, 2), (2, 3), (3, 2), (2, 4), (4, 2), (3. 3), (3, 4),

      (4, 3), and (4, 4).

The probability distribution for $\overline{X}$ is given below in the table.

$\overline{X}$	1	1.5	2	2.5	3	3.5	4
$P\{X\}$	0.01	0.04	0.10	0.20	0.25	0.24	0.16

Note, $P\{\overline{X}=1\} = P\{(1, 1)\} = P(1)\times P(1) = (0.1)(0.1) = 0.01$. Similarly for the other probabilities in the table.

(d) $E[\overline{X}] = 1(0.01) + 1.5(0.04) + ... + 4(0.16) = 3$.

(e) $Var(\overline{X}) = E[(\overline{X} - E[\overline{X}])^2\} = (1 - 3)^2(0.01) + (1.5 - 3)^2(0.04) + ... +$

$$(4-3)^2(0.16) \approx 0.5.$$

(f) $SD(\overline{X}) = \sqrt{0.5} = 0.707$.

5. Let X = number of left-handed people.

n = 100, p = 0.12.

(a) $\mu = np = 100(0.12) = 12$.

(b) $\sigma^2 = np(1 - p) = 12(1 - 0.12) = 10.56$.

(c) $\sigma = 3.2496$.

(d) $P\{10 \leq X \leq 14\} = P\{9.5 \leq X \leq 14.5\}$

$\approx P\{(9.5 - 12)/3.2496 \leq Z \leq (14.5 - 12)/3.2496\}$

$= P\{-0.7693 \leq Z \leq 0.7693\} = 2(0.7714 - 0.5) = 0.5588$.

7. Let X = sum of the monthly telephone bills.

$\mu = 15$, $\sigma = 7$, $n = 20$.

(a) $E[X] = n\mu = 20(15) = 300$.

(b) $Var(X) = (49)(20) = 980$.

   $SD(X) = \sqrt{(980)} = 31.3$.

(c) $P\{X > 300\} \approx P\{Z > (300 - 300)/2.45\} = P\{Z > 0\} = 0.5$.

9. Let X = number of consumers who were familiar with the product.

$n = 1000$, $p = 0.25$, $X = 232$, $np = 250$, $np(1 - p) = 187.5$,

$\sqrt{np(1-p)} = 13.69$.

$P\{X \le 232\} = P\{X \le 232.5\} \approx P\{Z \le (232.5 - 250)/13.69\}$

$= P\{Z \le -1.2783\} = 1 - 0.8997 = 0.1003$.

11. X is a binomial random variable with parameters $n = 80$ and $p = 0.4$.

$np = 32$, $\sqrt{np(1-p)} = 4.3818$.

(a) $P\{X > 34\} = P\{X \ge 34.5\} \approx P\{Z \ge (34.5 - 32)/4.3818\}$

   $= P\{Z \ge 0.5705\} = 1 - 0.7157 = 0.2843$.

(b) $P\{X \le 42\} = P\{X \le 42.5\} \approx P\{Z \le (42.5 - 32)/4.3818\}$

   $= P\{Z \le 2.396\} = 0.9918$.

(c) $P\{25 \leq X \leq 39\} = P\{24.5 \leq X \leq 39.5\}$

   $\approx P\{-1.7116 \leq Z \leq 1.7116\}$

   $= 0.9564 - (1 - 0.9564) = 0.9128$

13. (a) Let X = number of U.S. residents under 18 years of age who were not covered by health insurance in 2002.

   $n = 1000, X = 100, p = 0.116, np = 116, \sqrt{np(1-p)} = 10.1264.$

   $P\{X \geq 100\} = P\{X \geq 99.5\} \approx P\{Z \geq (99.5 - 116)/10.1264\}$

   $= P\{Z \geq -1.6294\} = P\{Z \leq 1.6294\} = 0.9484.$

   (b) Let X = number of U.S. residents who were between 25 and 34 years of age who were not covered by health insurance in 2002.

   $n = 1000, X = 260, p = 0.249, np = 249, \sqrt{np(1-p)} = 13.6748.$

   $P\{X < 260\} = P\{X \leq 259\} = P\{X \leq 259.5\}$

   $\approx P\{Z \leq (259.5 - 249)/13.6748\}$

   $= P\{Z \leq 0.7678\} = 0.7794.$

   (c) Let X = number of U.S. residents who were 65 and over years of age who were not covered by health insurance in 2002.
   Let Y = number of U.S. residents who were between 45 and 64 years of age who were not covered by health insurance in 2002.

$n = 1000, X = 5, p = 0.008, np = 8, \sqrt{np(1-p)} = 2.8171.$

$n = 1000, Y = 120, p = 0.135, np = 135, \sqrt{np(1-p)} = 10.8062.$

$P\{X \le 5\} = P\{X \le 5.5\} \approx P\{Z \le (5.5 - 8)/2.8171\}$

$= P\{Z \le -0.8874\} = 1 - P\{Z \le 0.8874\} = 1 - 0.8133 = 0.1867.$

$P\{Y \le 120\} = P\{Y \le 120.5\} \approx P\{Z \le (120.5 - 135)/10.8062\}$

$= P\{Z \le -1.3418\} = 1 - P\{Z \le 1.3418\} = 1 - 0.9099 = 0.0901.$

Thus, $P\{X \le 5 \text{ and } Y \le 120\} = P\{X \le 5\} \times P\{Y \le 120\}$

$= 0.1867 \times 0.0901 = 0.0168.$

(d) Let X = number of U.S. residents who were between 18 and 24 years of age who were not covered by health insurance in 2002.

Let Y = number of U.S. residents who were between 25 and 34 years of age who were not covered by health insurance in 2002.

X: $n = 1000, p = 0.296, np = 269, \text{Var}(X) = 189.376.$

Y: $n = 1000, p = 0.249, np = 249, \text{Var}(Y) = 186.999.$

$\sqrt{\text{Var}(X) + \text{Var}(Y)} = 19.4004.$

Thus, $P\{X > Y\} = P\{X - Y > 0\} \approx P\{Z > (0 - (269 - 249))/19.4004\}$

$= P\{Z > -1.0309\} = 1 - P\{Z \le 1.0309\} = 1 - 0.8485 = 0.1515.$

Chapter 8   ESTIMATION

SECTION 8.2   POINT ESTIMATOR OF A POPULATION MEAN

PROBLEMS

1. Estimate of the average weight of the participants in the race is

   $\overline{X}$ = 145.5 (pounds).

3. Estimate of the average amount spent by all students at the University is

   $\overline{X}$ = \$363.8.

5. Estimate of the average lifetime of a laser is

   $\overline{X}$ = 6624/40 =165.6 hours.

7. $\sigma = 10$, SD($\overline{X}$) = 3.

   Since SD($\overline{X}$) = $\dfrac{\sigma}{\sqrt{n}}$, then substituting and solving for n gives,

   $n = 100/9 \approx 12$.

9. Estimate of the average size of all single family household in the city is

   $\overline{X}$ = [1(11) + 2(19) + 3(28) + 4(26) + 5(11) + 6(4) + 7(1)]/100

   $= 3.23$.

11. (a) $SE(\overline{X}) = \sigma / \sqrt{n}$

(b) $SE(\overline{X}) = (\sqrt{3})[\sigma / \sqrt{n}] = 1.7321[\sigma / \sqrt{n}]$.

Comparing the standard errors for $\overline{X}$ in (a) and (b), we see that the standard error in (b) is larger. Thus, the data in (a) will yield a more precise estimator for μ.

(c) Let n_1 be the sample size in part (a) and n_2 be associated with part (b). Then, we want $\sigma / \sqrt{n_1} = (\sqrt{3})[\sigma / \sqrt{n_2}] = \sigma / \sqrt{(n_2 / 3)}$. That is, we want $n_1 = n_2/3$, from which $n_2 = 3n_1$. Thus, to achieve the same precision in (b) of the estimator in (a), we would need three times the sample size in (a).

SECTION 8.3   POINT ESTIMATOR OF A POPULATION PROPORTION

PROBLEMS

1. Let X = number of North Americans in 1985 who felt that the Communist party will win a free election in the Soviet Union.

   X = 510, n = 1325. Hence $\hat{p}$ = 510/1325 = 0.3849.

3. Let X = number of members who are in favor of increasing the annual dues.

   (a) X = 13, n = 20. Hence the estimate of the proportion of the members who are in favor is $\hat{p}$ = 13/20 = 0.65.

   (b) Estimate of the standard error is SD($\hat{p}$) = $\sqrt{\dfrac{\hat{p}(1-\hat{p})}{n}}$ = 0.1067.

5. Let X = number of students who owned a less than 5-year old car.

   (a) X = 35, n = 85. Hence the estimate of the proportion of students who owned a less than 5-year old car is $\hat{p}$ = 35/85 = 0.4118.

   (b) Estimate of the standard error is SD($\hat{p}$) = $\sqrt{\dfrac{\hat{p}(1-\hat{p})}{n}}$ = 0.0534.

7. Let X = number of construction workers who are presently unemployed.

   (a) X = 122, n = 1000. Hence the estimate of the proportion of

   unemployed construction workers is $\hat{p}= 122/1000 = 0.122$.

   (b) Estimate for the standard error is $SD(\hat{p}) = \sqrt{\dfrac{\hat{p}(1-\hat{p})}{n}} = 0.0103$.

9. n = 1200, number of Hispanic Engineers = 28, number of African American engineers = 45, number of females = 104.

   (a) Estimate of the proportion of Hispanic American engineers

   = 28/1200 = 0.0233.

   (b) Estimate of the proportion of African American engineers

   = 45/1200 = 0.0375.

   (c) Estimate of the proportion of female engineers

   = 104/1200 = 0.0867.

11. Let X = number of teenagers who died as a result of a motor vehicle accident.

    (a) X = 98, n = 400. Hence the estimate of the proportion of

    teenagers who died as a result of a motor vehicle accident is

    $\hat{p} = 98/400 = 0.245$.

(b) Estimate for the standard error is $SD(\hat{p}) = \sqrt{\dfrac{\hat{p}(1-\hat{p})}{n}} = 0.0215$.

13. Los Angeles: $SD(\hat{p}) \le \dfrac{1}{2\sqrt{n}} = 0.0091$.

San Diego: $SD(\hat{p}) < \dfrac{1}{2\sqrt{n}} = 0.0158$.

Now $0.0158/0.0091 = 1.7375$. Thus part (c) is the most accurate.

SECTION 8.3.1   ESTIMATING THE PROBABILITY OF A SENSITIVE EVENT

PROBLEMS

1. n = 50, number of "yes" response = 32 or 64%.

    $\hat{p} = 1 - 2(1 - 0.64) = 1 - 2(0.36) = 0.28.$

3. (a) The price would be higher for small values of p since

    $(1 + p)/p == 1/p + 1$. The value of $1/p$ will be increase

    when p decreases and thus $\sqrt{\dfrac{1+p}{p}}$ will increase.

    (b) 3.3166; 1.7321; 1.4530. These values indicate that as p is

    increasing, the value of $\sqrt{\dfrac{1+p}{p}}$ is decreasing.

SECTION 8.4   ESTIMATING A POPULATION VARIANCE

PROBLEMS

1. Let X = number of weekly hours worked by college professors.

$n = 10$, $\overline{X} = 45.50$, $\sum(X - \overline{X})^2 = 3034.5$, SD(X) = 18.36.

3. Let X = number of tons of chemical produced daily by the plant.

$n = 10$, $\overline{X} = 799.70$, $\sum(X - \overline{X})^2 = 1738.1$, SD(X) = 13.90,

Var(X) = 193.12.

5. Let X = weights of the runners in the 2004 Boston Marathon.

$n = 8$, $\overline{X} = 145.5$, $\sum(X - \overline{X})^2 = 3168$, SD(X) = 21.27.

7. Let X = observed data value.

$n = 10$, $\mu = 104$, $\sum(X - \mu)^2 = 305$, $\sigma = 5.52$, $\sigma^2 = 30.5$.

9. Let X = waiting time in minutes of the patients at the medical clinic.

$n = 12$, $\overline{X} = 44.08$, $\sum(X - \overline{X})^2 = 1756.92$, SD(X) = 12.64.

11. Let X = size of a single-family household.

$n = 100$, $\overline{X} = 3.23$, $\sum(X - \overline{X})^2 = 179.71$, SD(X) = 1.347.

13. Let X = closing price of crude oil on the New York Mercantile
     Exchange.

     n = 20, $\overline{X}$ = 17.465, $\sum(X - \overline{X})^2$ = 3.4351, SD(X) = 0.4252.

SECTION 8.5 INTERVAL ESTIMATOR OF THE MEAN OF A NORMAL POPULATION WITH KNOWN POPULATION VARIANCE

PROBLEMS

1. Let X = reading of the scale.

   (a) $\sigma = 0.1$, $n = 5$, $\overline{X} = 3.1502$, $\alpha = 0.05$, $Z_{\alpha/2} = 1.96$.

   The 95% confidence interval for the true mean of the weight is

   $$\overline{X} \pm Z_{\alpha/2} \times \sigma/\sqrt{n} \Rightarrow 3.1502 \pm 0.0877 \Rightarrow (3.0625, 3.2379).$$

   (b) $\sigma = 0.1$, $n = 5$, $\overline{X} = 3.1502$, $\alpha = 0.01$, $Z_{\alpha/2} = 2.576$.

   The 99% confidence interval for the true mean of the weight is

   $$\overline{X} \pm Z_{\alpha/2} \times \sigma/\sqrt{n} \Rightarrow 3.1502 \pm 0.1152 \Rightarrow (3.035, 3.2654).$$

3. Let X = PCB level in the fish.

   $\sigma = 0.8638$, $n = 10$, $\overline{X} = 11.48$, $\alpha = 0.05$, $Z_{\alpha/2} = 1.96$.

   The 95% confidence interval for the true mean PCB level in the fish is

   $$\overline{X} \pm Z_{\alpha/2} \times \sigma/\sqrt{n} \Rightarrow 11.48 \pm 0.5354 \Rightarrow (10.9446, 12.0154).$$

5. Let X= life of the TV tube.

   (a) $\sigma = 400$, $n = 20$, $\overline{X} = 9000$, $\alpha = 0.1$, $Z_{\alpha/2} = 1.645$.

   The 90% confidence interval for the true mean life of the TV tube is

$\overline{X} \pm Z_{\alpha/2} \times \sigma/\sqrt{n} \Rightarrow 9000 \pm 147.1333 \Rightarrow (8852.8667, 9147.1333).$

(b) $\sigma = 400$, $n = 20$, $\overline{X} = 9000$, $\alpha = 0.1$, $Z_{\alpha/2} = 1.96$.

The 95% confidence interval for the true mean life of the TV tube is

$\overline{X} \pm Z_{\alpha/2} \times \sigma/\sqrt{n} \Rightarrow 9000 \pm 175.3077 \Rightarrow (8824.6923, 9175.3077).$

7. Let X = test score of a student.

$\sigma = 11.3$, $n = 81$, $\overline{X} = 74.6$, $\alpha = 0.1$, $Z_{\alpha/2} = 1.645$.

The 90% confidence interval for the true mean test score is

$\overline{X} \pm Z_{\alpha/2} \times \sigma/\sqrt{n} \Rightarrow 74.6 \pm 2.0654 \Rightarrow (72.5346, 76.6654).$

9. Let X = life of the lightbulb.

(a) $\sigma = 100$, $n = 169$, $\overline{X} = 1350$, $\alpha = 0.1$, $Z_{\alpha/2} = 1.645$.

The 90% confidence interval for the true mean life of the bulb is

$\overline{X} \pm Z_{\alpha/2} \times \sigma/\sqrt{n} \Rightarrow 1350 \pm 12.6538 \Rightarrow (1337.3462, 1336.6538).$

(b) $\sigma = 100$, $n = 169$, $\overline{X} = 1350$, $\alpha = 0.05$, $Z_{\alpha/2} = 1.96$.

The 95% confidence interval for the true mean life of the bulb is

$\overline{X} \pm Z_{\alpha/2} \times \sigma/\sqrt{n} \Rightarrow 1350 \pm 15.0769 \Rightarrow (1334.9231, 1365.0769).$

(c) $\sigma = 100$, $n = 169$, $\overline{X} = 1350$, $\alpha = 0.01$, $Z_{\alpha/2} = 2.576$.

The 95% confidence interval for the true mean life of the bulb is

$$\overline{X} \pm Z_{\alpha/2} \times \sigma/\sqrt{n} \Rightarrow 1350 \pm 19.8154 \Rightarrow (1330.1846, 1369.8154).$$

11. $\sigma = 3300$, $\alpha = 0.01$, $Z_{\alpha/2} = 2.576$, $b = 4090.7224$ (from problem 10).

$$n \ge [(2 \times Z_{\alpha/2} \times \sigma)/b]^2 = 17.2735 \approx 18.$$

13. $\sigma = 180$, $\alpha = 0.05$, $Z_{\alpha/2} = 1.96$, $b = 40$.

$$n \ge [(2 \times Z_{\alpha/2} \times \sigma)/b]^2 = 311.1696 \approx 312.$$

15. Let X = test score of a student.

   (a) $\sigma = 11.3$, $n = 81$, $\overline{X} = 74.6$, $\alpha = 0.1$, $Z_\alpha = 1.28$.

   The 90% lower confidence bound for the true mean test score is

   $$\overline{X} - Z_{\alpha/2} \times \sigma/\sqrt{n} \Rightarrow 74.6 - 1.6071 = 72.9929.$$

   (b) $\sigma = 11.3$, $n = 81$, $\overline{X} = 74.6$, $\alpha = 0.05$, $Z_\alpha = 1.645$.

   The 95% lower confidence bound for the true mean test score is

   $$\overline{X} - Z_{\alpha/2} \times \sigma/\sqrt{n} \Rightarrow 74.6 - 2.0654 = 72.5346.$$

   (c) $\sigma = 11.3$, $n = 81$, $\overline{X} = 74.6$, $\alpha = 0.05$, $Z_\alpha = 1.645$.

   The 95% upper confidence bound for the true mean test score is

   $$\overline{X} + Z_{\alpha/2} \times \sigma/\sqrt{n} \Rightarrow 74.6 + 2.0654 = 76.6654.$$

   (d) $\sigma = 11.3$, $n = 81$, $\overline{X} = 74.6$, $\alpha = 0.01$, $Z_\alpha = 2.33$.

   The 95% upper confidence bound for the true mean test score is

$$\overline{X} + Z_{\alpha/2} \times \sigma/\sqrt{n} \Rightarrow 74.6 + 2.9254 = 77.5254.$$

17. For problem number 10, the 95% confidence interval for the average tire life is (26,354.6388, 30,445.3612). The advertisement will not be false, however it is not revealing the entire truth.

SECTION 8.6   *INTERVAL ESTIMA TORS OF THE MEAN*
OF A NORMAL POPULATION WITH
UNKNOWN POPULATION VARIANCE

PROBLEMS

1. Let X = time (years) for graduate student to complete his/her degree.

   (a) $S = 1.2$, $n = 2000$, $\overline{X} = 5.2$, $\alpha = 0.05$, $t_{\alpha/2, \, 1999} \approx Z_{\alpha/2} = 1.96$.

   The 95% confidence interval for the true mean completion time is

   $$\overline{X} + Z_{\alpha/2} \times S/\sqrt{n} \implies 5.2 \pm 0.0526 \implies (5.1474, \, 5.2526).$$

   (b) $S = 1.2$, $n = 2000$, $\overline{X} = 5.2$, $\alpha = 0.01$, $t_{\alpha/2, \, 1999} \approx Z_{\alpha/2} = 2.576$.

   The 99% confidence interval for the true mean completion time is

   $$\overline{X} + Z_{\alpha/2} \times S/\sqrt{n} \implies 5.2 \pm 0.0691 \implies (5.1309, \, 5.2691).$$

3. Let $\dot{X}$ = salary of a middle-level bank executive.

   (a) $S = 22.0871$, $n = 15$, $\overline{X} = 83.8667$, $\alpha = 0.1$, $t_{\alpha/2, \, 14} = 1.761$.

   The 90% confidence interval for the true mean salary of a

   middle-level bank executive is

   $$\overline{X} + t_{\alpha/2, n-1} \times S/\sqrt{n} \implies 83.8667 \pm 10.0427 \implies (73.8240, \, 93.9094).$$

(b) $S = 22.0871$, $n = 15$, $\overline{X} = 83.8667$, $\alpha = 0.05$, $t_{\alpha/2,\,14} = 2.145$.

The 95% confidence interval for the true mean salary of a

middle-level bank executive is

$$\overline{X} + t_{\alpha/2,n-1} \times S/\sqrt{n} \Rightarrow 83.8667 \pm 12.2326 \Rightarrow (71.6341,\ 96.0993).$$

(c) $S = 22.0871$, $n = 15$, $\overline{X} = 83.8667$, $\alpha = 0.01$, $t_{\alpha/2,\,14} = 2.977$.

The 99% confidence interval for the true mean salary of a

middle-level bank executive is

$$\overline{X} + t_{\alpha/2,n-1} \times S/\sqrt{n} \Rightarrow 83.8667 \pm 16.9774 \Rightarrow (66.8893,\ 100.8441).$$

5. Let X = weight of a participant in the 2004 Boston Marathon.

(a) $S = 21.2737$, $n = 8$, $\overline{X} = 145.5$, $\alpha = 0.05$, $t_{\alpha/2,\,7} = 2.365$.

The 95% confidence interval for the true mean weight of the

Boston Marathon participants in 2004 is

$$\overline{X} + t_{\alpha/2,n-1} \times S/\sqrt{n} \Rightarrow 145.5 \pm 17.7881 \Rightarrow (127.7119,\ 163.2881).$$

(b) $S = 21.2737$, $n = 8$, $\overline{X} = 145.5$, $a = 0.01$, $t_{\alpha/2,\,7} = 3.499$.

The 99% confidence interval for the true mean weight of the

Boston Marathon participants in 2004 is

$$\overline{X} + t_{\alpha/2,n-1} \times S/\sqrt{n} \Rightarrow 145.5 \pm 26.3173 \Rightarrow (119.1827,\ 171.8173).$$

7. Let X = burn time for a chair.

    $S = 19.32, n = 7, \overline{X} = 464.14, \alpha = 0.05, t_{\alpha/2, 6} = 2.447.$

    The 95% confidence interval for the true mean bum temperature is

    $\overline{X} + t_{\alpha/2, n-1} \times S/\sqrt{n} \Rightarrow 464.14 \pm 17.8687 \Rightarrow (446.2713, 482.0087).$

9. Let X = a winning score in the Masters Golf Tournament.

    $S = 3.6645, n = 8, \overline{X} = 282.50, \alpha = 0.1, t_{\alpha/2, 7} = 1.895.$

    The 90% confidence interval for the true mean winning score is

    $\overline{X} + t_{\alpha/2, n-1} \times S/\sqrt{n} \Rightarrow 282.50 \pm 2.4552 \Rightarrow (280.0448, 284.9552).$

11. Let X = repair cost for one of the cars.

    $S = 800, n = 16, \overline{X} = 2200, \alpha = 0.1, t_{\alpha/2, 15} = 1.753.$

    The 90% confidence interval for the true mean winning score is

    $\overline{X} + t_{\alpha/2, n-1} \times S/\sqrt{n} \Rightarrow 2200 \pm 350.6 \Rightarrow (1849.4, 2550.6).$

13. Let X = length of telephone call made during midday.

    (a) $S = 2.2, n = 1200, \overline{X} = 4.7, \alpha = 0.1, t_{\alpha/2, 1199} \approx Z_{\alpha/2} = 1.645.$

    The 90% confidence interval for the true mean time for a telephone call made during midday is

    $\overline{X} + Z_{\alpha/2} \times S/\sqrt{n} \Rightarrow 4.7 \pm 0.1045 \Rightarrow (4.5955, 4.8045).$

(b) $S = 2.2$, $n = 1200$, $\overline{X} = 4.7$, $\alpha = 0.05$, $t_{\alpha/2, 1199} \approx Z_{\alpha/2} = 1.96$.

The 95% confidence interval for the true mean time for a

telephone call made during midday is

$$\overline{X} + Z_{\alpha/2} \times S/\sqrt{n} \Rightarrow 4.7 \pm 0.1245 \Rightarrow (4.5755, 4.8245).$$

15. Let X = debt on a CitiBank VISA card.

$S = 840$, $n = 300$, $\overline{X} = 1220$, $\alpha = 0.05$, $t_{\alpha/2, 299} \approx Z_{\alpha/2} = 1.96$.

The 95% confidence interval for the true mean debt on a CitiBank

VISA card is

$$\overline{X} + Z_{\alpha/2} \times S/\sqrt{n} \Rightarrow 1220 \pm 95.0549 \Rightarrow (1124.9451, 1315.0549).$$

17. No. We can say that in the long run, if repeated samples of
size 9 are selected from the population, then we can expect that 95%
of the intervals to contain the true population mean. If a single
sample of size 9 is selected and the 95% confidence interval is
constructed, then the population mean will either be contained in the
interval or not so we cannot be 95% "certain".

19. Let X = lifetime of a lightbulb.

(a) $S = 8.6989$, $n = 12$, $\overline{X} = 33.1167$, $\alpha = 0.05$, $t_{\alpha/2, 11} = 2.201$.

The 95% confidence interval for the true mean lifetime of the

bulbs is

$$\overline{X} + t_{\alpha/2,n-1} \times S/\sqrt{n} \Rightarrow 33.1167 \pm 5.5271 \Rightarrow (27.5896, 38.6438).$$

(b) $S = 8.6989$, $n = 12$, $\overline{X} = 33.1167$, $\alpha = 0.01$, $t_{\alpha/2, 11} = 3.106$.

The 99% confidence interval for the true mean lifetime of the bulbs

is

$$\overline{X} + t_{\alpha/2, n-1} \times S/\sqrt{n} \Rightarrow 33.1167 \pm 7.7997 \Rightarrow (25.317, 40.9164).$$

No, you cannot make the claim. The interval indicates that the

mean lifetime could also be less than 30 hours.

21. Let X = salary of a middle-level bank executive.

(a) $S = 22.0871$, $n = 15$, $\overline{X} = 83.8667$, $\alpha = 0.01$, $t_{\alpha, 14} = 2.624$.

The 99% upper confidence bound for the true mean salary of a

middle-level bank executive is

$$\overline{X} + t_{\alpha/2, n-1} \times S/\sqrt{n} \Rightarrow 83.8667 + 14.9643 = 98.8310. \text{ Thus, an}$$

appropriate value for v_1 is 98.8310.

(b) $S = 22.0871$, $n = 15$, $\overline{X} = 83.8667$, $\alpha = 0.01$, $t_{\alpha, 14} = 2.624$.

The 99% lower confidence bound for the true mean salary of a

middle-level bank executive is

$$\overline{X} - t_{\alpha/2, n-1} \times S/\sqrt{n} \Rightarrow 83.8667 - 14.9643 = 68.9024. \text{ Thus, an}$$

appropriate value for v_2 is 68.9024.

23. The executive should present a lower confidence bound for the average daily cash receipts.

Let X = daily cash receipt.

$S = 11.6187$, $n = 14$, $\overline{X} = 34.0714$, $\alpha = 0.05$, $t_{\alpha, 13} = 1.771$.

The 95% lower confidence bound for the true mean daily cash receipt is

$\overline{X} - t_{\alpha/2, n-1} \times S/\sqrt{n} \Rightarrow 34.0714 - 5.4994 = 28.572$. Thus, "I am 95% confident that the mean daily receipt is at least \$2,587.20".

SECTION 8.7   INTERVAL ESTIMATORS OF A POPULATION PROPORTION

PROBLEMS

1. Let X = number of voters in the sample who favor the death penalty.

    $n = 500$, $X = 302$, $\hat{p} = X/n = 302/500 = 0.604$, $\alpha = 0.01$, $Z_{\alpha/2} = 2.576$.

    The 99% confidence interval for the true proportion of California

    voters who favor the death penalty is

    $$\hat{p} \pm Z_{\alpha/2}\sqrt{\frac{\hat{p}(1-\hat{p})}{n}} \Rightarrow 0.604 \pm 0.0563 \Rightarrow (0.5477, 0.6603).$$

3. Let X = number of male babies in the sample.

    (a) $n = 10,000$, $X = 5106$, $\hat{p} = X/n = 5106/1000 = 0.5106$, $\alpha = 0.1$,

    $Z_{\alpha/2} = 1.645$.

    The 90% confidence interval for the true proportion of male babies

    $$\hat{p} \pm Z_{\alpha/2}\sqrt{\frac{\hat{p}(1-\hat{p})}{n}} \Rightarrow 0.5106 \pm 0.0082 \Rightarrow (0.5024, 0.5188).$$

    (b) $n = 10,000$, $X = 5106$, $\hat{p} = X/n = 5106/1000 = 0.5106$, $\alpha = 0.01$,

    $Z_{\alpha/2} = 2.576$.

    The 90% confidence interval for the true proportion of male babies

$$\hat{p} \pm Z_{\alpha/2} \sqrt{\frac{\hat{p}(1-\hat{p})}{n}} \Rightarrow 0.5106 \pm 0.0129 \Rightarrow (0.4977, 0.5235).$$

5. Let X = number of Los Angeles residents who are in favor of strict gun control legislation in the sample.

   n = 100, X = 64, $\hat{p}$ = X/n = 64/100 = 0.64, α = 0.05, $Z_{\alpha/2}$ = 1.96.

   The 95% confidence interval for the true proportion of Los Angeles

   residents who favor strict gun control legislation is

$$\hat{p} \pm Z_{\alpha/2} \sqrt{\frac{\hat{p}(1-\hat{p})}{n}} \Rightarrow 0.64 \pm 0.0941 \Rightarrow (0.5459, 0.7341).$$

7. Let X = number of North Americans in the sample who felt that the Communist party would have won a free election in the Soviet Union.

   X = 510, n = 1325, $\hat{p}$ = 510/1325 = 0.3849, α = 0.05, $Z_{\alpha/2}$ = 1.96.

   The 95% confidence interval for the true proportion of North Americans who felt that the Communist party would have won a free election in the Soviet Union is

$$\hat{p} \pm Z_{\alpha/2} \sqrt{\frac{\hat{p}(1-\hat{p})}{n}} \Rightarrow 0.3849 \pm 0.0262 \Rightarrow (0.3587, 0.4111).$$

9. Let X = number of bottles of wine in the sample that are spoiled.

   X = 3, n = 20, $\hat{p}$ = 3/20. The normal approximation does not hold since

   $n\hat{p} = (20)(3/20) = 3 < 5.$

11. Let X = number of female librarians in the sample.

X = 335, n = 400, $\hat{p} = 335/400 = 0.8375$, $\alpha = 0.05$, $Z_{\alpha/2} = 1.96$.

The 95% confidence interval for the true proportion of female

librarians is

$$\hat{p} \pm Z_{\alpha/2}\sqrt{\frac{\hat{p}(1-\hat{p})}{n}} \Rightarrow 0.8375 \pm 0.0362 \Rightarrow (0.8013, 0.8037).$$

13. Let X = number of states that had a 1990 per capita income in excess of $20,000 in the sample.

X = 2, n = 9, $\hat{p} = 2/9$. The normal approximation does not hold since

$n\hat{p} = (9)(2/9) = 2 < 5$.

15. (a) Let X = number of African Americans in the sample.

X = 42, n = 500, $\hat{p} = 42/500 = 0.084$, $\alpha = 0.05$, $Z_{\alpha/2} = 1.96$.

The 95% confidence interval for the true proportion of African

American accountants is

$$\hat{p} \pm Z_{\alpha/2}\sqrt{\frac{\hat{p}(1-\hat{p})}{n}} \Rightarrow 0.084 \pm 0.0243 \Rightarrow (0.0597, 0.1083).$$

(b) Let X = number of Hispanic Americans in the sample.

$$X = 18, n = 500, \hat{p} = 18/500 = 0.036, \alpha = 0.05, Z_{\alpha/2} = 1.96.$$

The 95% confidence interval for the true proportion of Hispanic American accountants is

$$\hat{p} \pm Z_{\alpha/2} \sqrt{\frac{\hat{p}(1-\hat{p})}{n}} \Rightarrow 0.036 \pm 0.0163 \Rightarrow (0.0197, 0.0523).$$

(c) Let X = number of females in the sample.

$$X = 246, n = 500, \hat{p} = 246/500 = 0.492, \alpha = 0.05, Z_{\alpha/2} = 1.96.$$

The 95% confidence interval for the true proportion of female accountants is

$$\hat{p} \pm Z_{\alpha/2} \sqrt{\frac{\hat{p}(1-\hat{p})}{n}} \Rightarrow 0.492 \pm 0.0438 \Rightarrow (0.4482, 0.5358).$$

17. (a) A 95% confidence interval is given by 0.75 ± 0.0346.

(b) Perhaps a 98% confidence interval was used.

19. (a) $\alpha = 0.1, Z_{\alpha/2} = 1.645, b = 0.04.$

Since $n > (Z_{\alpha/2}/b)^2$, then $n > (1.645/0.04)^2 = 1691.2656 \approx 1692.$

(b) $n = 1692, \hat{p} = 0.23, \alpha = 0.1, Z_{\alpha/2} = 1.645.$

The 90% confidence interval for the true proportion of households that are watching a particular sporting event is

$$\hat{p} \pm Z_{\alpha/2}\sqrt{\frac{\hat{p}(1-\hat{p})}{n}} \Rightarrow 0.23 \pm 0.0168 \Rightarrow (0.2132, 0.2468). \text{ Thus}$$

the length of the confidence interval is 0.2468 - 0.2132 = 0.0336.

That is, the length of the interval will be smaller than 0.04(2×0.02).

(c) The 90% confidence interval is given in part (b).

21. Let L be the length of the interval. Thus, L = 0.02 and $\hat{p}$ = 0.2. Now,

$$n = 4 \times Z_{\alpha/2}^2 \times \frac{p(1-p)}{L^2} = (4)(1.96)^2(0.2)(0.8)/(0.02)^2 = 6146.56 \approx 6147.$$

23. Let X = number of female librarians in the sample.

X = 335, n = 400, $\hat{p}$ = 335/400 = 0.8375, α = 0.05, Z_{α} = 1.645.

The 95% upper confidence interval for the true proportion of female

librarians is

$$\hat{p} + Z_{\alpha}\sqrt{\frac{\hat{p}(1-\hat{p})}{n}} \Rightarrow 0.8375 + 0.0304 = 0.8679.$$

25. (a) Let X = number of African Americans or Hispanic Americans in the
         sample.

X = 60, n = 500, $\hat{p}$ = 60/500 = 0.12, α = 0.1, Z_{α} = 1.28.

The 90% lower confidence interval for the true proportion of
African American accountants is

$$\hat{p} - Z_\alpha \sqrt{\frac{\hat{p}(1-\hat{p})}{n}} \Rightarrow 0.12 - 0.0186 = 0.1014.$$

(b) Let X = number of African Americans or Hispanic Americans in the sample.

X = 60, n=500, $\hat{p}$ = 60/500 = 0.12, α = 0.1, Z_α = 1.28.

The 90% upper confidence interval for the true proportion of

African Americans or Hispanic American accountants is

$$\hat{p} - Z_\alpha \sqrt{\frac{\hat{p}(1-\hat{p})}{n}} \Rightarrow 0.12 + 0.0186 = 0.1386.$$

27. (a) Let X = number of bottles of wine in the sample that are spoiled.

X = 3, n = 20, $\hat{p}$ = 3/20. The normal approximation does not hold

since $n\hat{p}$ = (20)(3/20) = 3 < 5.

(b) The 99% confidence interval will be (0, 0.3557). Thus, we have the same conclusion as in part (a).

REVIEW PROBLEMS

1. For part (a), $SD(\overline{X}) = \sigma/2\sqrt{n}$ and for part (b), $SD(\overline{X}) = \sigma/\sqrt{2n}$. Since

    $SD(\overline{X})$ for part (a) is smaller, then the situation in part (a) will produce a

    more precise estimator for μ.

3. Let X = amount of time (hours) a person spent watching TV in a given
    week.

    $S = 7.4$, $n = 50$, $\overline{X} = 24.4$, $\alpha = 0.05$, $t_{\alpha/2, 49} \approx Z_{\alpha/2} = 1.96$.

    The 95% confidence interval for the true mean amount of time a

    person spends watching TV in a given week is

    $\overline{X} + Z_{\alpha/2} \times S/\sqrt{n} \Rightarrow 24.4 \pm 2.0512 \Rightarrow (22.3488, 26.4512)$.

5. Let X = score of a 6th-grade student in the state of Washington

    $S = 16$, $n = 100$, $\overline{X} = 320$, $\alpha = 0.05$, $t_{\alpha/2, 99} \approx Z_{\alpha/2} = 1.96$.

    The 95% confidence interval for the true mean score for the students
    is
    $\overline{X} + Z_{\alpha/2} \times S/\sqrt{n} \Rightarrow 320 \pm 3.136 \Rightarrow (316.864, 323.136)$.

7. Let X = number of drinks sold from a vending machine on a given day.

    (a) $n = 20$, $\overline{X} = 49.6$, $S = 10.1794$, $\alpha = 0.05$, $t_{\alpha/2, 19} = 2.093$.

    The 95% confidence interval for the true mean number of drinks
    sold daily is

    $\overline{X} + t_{\alpha/2, 19} \times S/\sqrt{n} \Rightarrow 49.6 \pm 4.7641 \Rightarrow (44.8359, 54.3641)$.

(b) $n = 20$, $\overline{X} = 49.6$, $S = 10.1794$, $\alpha = 0.1$, $t_{\alpha/2,\,19} = 1.734$.

The 90% confidence interval for the true mean number of drinks sold daily is

$$\overline{X} + t_{\alpha/2,\,19} \times S/\sqrt{n} \Rightarrow 49.6 \pm 3.9469 \Rightarrow (45.6531, 53.5469).$$

9. Let X = cost per delivery.

$n = 24$, $\overline{X} = 1840$, $S = 740$, $\alpha = 0.05$, $t_{\alpha/2,\,23} = 2.069$.

The 95% confidence interval for the true mean cost per delivery is

$$\overline{X} + t_{\alpha/2,\,23} \times S/\sqrt{n} \Rightarrow 1840 \pm 312.5263 \Rightarrow (1527.47, 2152.53).$$

11. Let X = speed of one of the pitcher's fastball.

(a) The point estimate for the mean speed of the pitcher's fastball is

$\overline{X} = 88.5556$ miles per hour.

(b) $n = 9$, $\overline{X} = 88.5556$, $S = 7.1609$, $\alpha = 0.05$, $t_{\alpha/2,\,8} = 2.306$.

The 95% confidence interval for the true mean speed of the pitcher's fast ball is

$$\overline{X} + t_{\alpha/2,\,8} \times S/\sqrt{n} \Rightarrow 88.5556 \pm 5.5043 \Rightarrow (83.0513, 94.0599).$$

13. Let X = value of an observation.

(a) $\sigma = 3$, $n = 36$, $\overline{X} = 35$, $\alpha = 0.05$, $Z_{\alpha/2} = 1.96$.

The 95% confidence interval for the population mean is

$$\overline{X} + Z_{\alpha/2} \times \sigma/\sqrt{n} \Rightarrow 35 \pm 0.98 \Rightarrow (34.02, 35.98).$$

(b) $\sigma = 6$, $n = 36$, $\overline{X} = 35$, $\alpha = 0.05$, $Z_{\alpha/2} = 1.96$.

The 95% confidence interval for the population mean is

$$\overline{X} + Z_{\alpha/2} \times \sigma/\sqrt{n} \Rightarrow 35 \pm 1.96 \Rightarrow (33.04, 36.96).$$

(c) $\sigma = 12$, $n = 36$, $\overline{X} = 35$, $\alpha = 0.05$, $Z_{\alpha/2} = 1.96$.

The 95% confidence interval for the population mean is

$$\overline{X} + Z_{\alpha/2} \times \sigma/\sqrt{n} \Rightarrow 35 \pm 3.92 \Rightarrow (31.08, 38.92).$$

15. Let X = number of female secondary school teachers.

$X = 518$, $n = 1000$, $\hat{p} = 518/1000 = 0.518$, $\alpha = 0.05$, $Z_{\alpha/2} = 1.96$.

The 95% confidence interval for the true proportion of secondary school teachers who are females is

$$\hat{p} \pm Z_{\alpha/2} \sqrt{\frac{\hat{p}(1-\hat{p})}{n}} \Rightarrow 0.518 \pm 0.0310 \Rightarrow (0.4870, 0.5490).$$

17. The largest possible margin of error is computed from

$$\pm Z_{\alpha/2} \sqrt{\frac{\hat{p}(1-\hat{p})}{n}}$$ when $p = 1/2$. Thus the largest possible margin of error is $\pm \dfrac{Z_{\alpha/2}}{2\sqrt{50,000}}$.

19. (a) Let X = number of runs (at least one) scored when there is a
       player on first base and there are no outs.

    n = 1728, $\hat{p}$ = 0.396, α = 0.05, $Z_{\alpha/2}$ = 1.96.

    The 95% confidence interval for the true proportion of at least one
    run being scored for this situation is

    $$\hat{p} \pm Z_{\alpha/2} \sqrt{\frac{\hat{p}(1-\hat{p})}{n}} \Rightarrow 0.396 \pm 0.0231 \Rightarrow (0.3729, 0.4191).$$

    (b) Let X = number of runs (at least one) scored when there is a player
       on second base and there is one out.

    n = 657, $\hat{p}$ = 0.39, α = 0.05, $Z_{\alpha/2}$ = 1.96.

    The 95% confidence interval for the true proportion of at least one
    run being scored for this situation is

    $$\hat{p} \pm Z_{\alpha/2} \sqrt{\frac{\hat{p}(1-\hat{p})}{n}} \Rightarrow 0.39 \pm 0.0373 \Rightarrow (0.3527, 0.4273).$$

21. Let X = number of secondary school teachers that are female.
    (a) n = 1000, X = 518, $\hat{p}$ = 518/1000 = 0.518, α = 0.1, $Z_{0.1}$ = 1.28.

    The 90% upper confidence bound for the proportion of female

    secondary school teachers is

    $$\hat{p} + Z_{\alpha} \sqrt{\frac{\hat{p}(1-\hat{p})}{n}} \Rightarrow 0.518 + (1.28)(0.0158) = 0.5382.$$

The lower confidence bound is 0.518 - (1.28)(0.0158) = 0.4978.

(b) n = 1000, X = 518, $\hat{p}$ = 518/1000 = 0.518, a = 0.05, $Z_{0.05}$ = 1.645.

The 95% upper confidence bound is 0.5440. Similarly, the lower confidence bound is 0.4920.

(c) n = 1000, X = 518, $\hat{p}$ = 518/1000 = 0.518, α = 0.01, $Z_{0.01}$ = 2.33.

The 99% upper confidence bound is 0.5548. The lower confidence bound is 0.4812.

Based on these bounds, it would be "wise" to quote the upper bounds.

Chapter 9   TESTING STATISTICAL HYPOTHESES

SECTION 9.2   HYPOTHESIS TESTS AND SIGNIFICANCE LEVELS

PROBLEMS

1. (a) Hypothesis B.

   (b) The appropriate level of significance (probability of a Type I error) in this situation should be zero.

3. Statement (d).

SECTION 9.3   TESTS CONCERNING THE MEAN OF A NORMAL POPULATION: CASE OF KNOWN VARIANCE

PROBLEMS

1. $n = 6$, $\overline{X} = 14.7167$, $\sigma = 0.5$, $SE(\overline{X}) = 0.2041$, $\alpha = 0.05$, $Z_{\alpha/2} = 1.96$.

H_0: $\mu = 14.4$ against H_1: $\mu \neq 14.4$

Test statistic: $Z = 1.5517$

Conclusion: Since $|Z| = 1.5517 < Z_{\alpha/2} = 1.96$, do not reject H_0. That is, there is insufficient sample evidence to conclude that the average distance (light years) from the earth to the asteroid phyla is not equal to 14.4 light years at the 5% level of significance.

3. (a) $n = 9$, $\overline{X} = 100$, $\sigma = 5$, $SE(\overline{X}) = 1.6667$, $\alpha = 0.05$, $Z_{\alpha/2} = 1.96$.

H_0: $\mu = 105$ against H_1: $\mu \neq 105$

Test statistic: $Z = -3$

p-value $= 2P\{Z > 3.0\} = 2(1 - 0.9987) = 0.0026$.

Conclusion: Since $|Z| = 3 > Z_{\alpha/2} = 1.96$, reject H_0.

Note: When $\alpha = 0.01$, $Z_{\alpha/2} = 2.575$, so we will still reject H_0. This conclusion is also supported by the p-value.

(b) $n = 9$, $\overline{X} = 100$, $\sigma = 10$, $SE(\overline{X}) = 3.3333$, $\alpha = 0.05$, $Z_{\alpha/2} = 1.96$.

H_0: $\mu = 105$ against H_1: $\mu \neq 105$

Test statistic: $Z = -1.5$

p-value $= 2P\{Z > 1.5\} = 2(1 - 0.9332) = 0.1336$.

Conclusion: Since $|Z| = 1.5 < Z_{\alpha/2} = 1.96$, do not reject H_0.

Note: When $\alpha = 0.01$, $Z_{\alpha/2} = 2.575$, so we still will not reject H_0. This conclusion is also supported by the p-value.

(c) $n = 9$, $\overline{X} = 100$, $\sigma = 15$, $SE(\overline{X}) = 5$, $\alpha = 0.05$, $Z_{\alpha/2} = 1.96$.

H_0: $\mu = 105$ against H_1: $\mu \neq 105$

Test statistic: $Z = -1.0$

p-value $= 2P\{Z > 1\} = 2(1 - 0.8413) = 0.3174$.

Conclusion: Since $|Z| = 1 < Z_{\alpha/2} = 1.96$, do not reject H_0.

Note: When $\alpha = 0.01$, $Z_{\alpha/2} = 2.575$, so we still will not reject H_0. This conclusion is also supported by the p-value.

5. $n = 25$, $\overline{X} = 30.4$, $\sigma = 4$, $SE(\overline{X}) = 0.8$, $\alpha = 0.05$, $Z_{\alpha/2} = 1.96$.

H_0: $\mu = 32$ against H_1: $\mu \neq 32$

Test statistic: $Z = -2.0$

p-value $= 2P\{Z > 2.0\} = 2(1 - 0.9772) = 0.0456$.

Conclusion: Since $|Z| = 2 > Z_{\alpha/2} = 1.96$, reject H_0.
That is, there is sufficient sample evidence to conclude that
the average weight of the mice is not equal to 32 grams at the
5% level of significance. This may indicate some bias in the
sampling procedure.

7. (a) $n = 200$, $\overline{X} = 374$, $\sigma = 40$, $SE(\overline{X}) = 2.8284$, $\alpha = 0.05$,
$Z_{\alpha/2} = 1.96$.

H_0: $\mu = 360$ against H_1: $\mu \neq 360$

Test statistic: $Z = 4.9498$

Conclusion: Since $|Z| = 4.9498 > Z_{\alpha/2} = 1.96$, reject H_0. That is,
there is enough sample evidence to conclude that the average
daily water usage is not equal to 360 gallons at the 5% level of
significance.

(b) p-value $= 2P\{Z > 4.9498\} \approx 0.0$.

9. $n = 36$, $\overline{X} = 15{,}233$, $\sigma = 4{,}000$, $SE(\overline{X}) = 666.6667$, $\alpha = 0.05$,
$Z_{\alpha/2} = 1.96$.

H_0: $\mu = 13{,}500$ against H_1: $\mu \neq 13{,}500$

Test statistic: $Z = 2.5995$

p-value $= 2P\{Z > 2.5995\} = 2(1 - 0.9953) = 0.0094$.

Conclusion: Since $|Z| = 2.5995 > Z_{\alpha/2} = 1.96$ or since the p-value $= 0.0094 < 0.05$, reject H_0. That is, there is enough sample evidence to conclude that the average miles driven by a one-year-old leased car is not equal to 13,500 miles at the 5% level of significance.

11. $n = 40$, $\overline{X} = 32.2$, $\sigma = 1.4$, $SE(\overline{X}) = 0.2214$, $\alpha = 0.05$, $Z_{\alpha/2} = 1.96$.

   H_0: $\mu = 30$ against H_1: $\mu \neq 30$

   Test statistic: $Z = 9.9368$

   p-value $= 2P\{Z > 9.9368\} \approx 0.0$.

   Conclusion: Since $|Z| = 9.9368 > Z_{\alpha/2} = 1.96$ or since the p-value $= 0.0 < 0.05$, reject H_0. That is, there is enough sample evidence to conclude that the average duration of a red light is not equal to 30 seconds at the 5% level of significance.

   Same conclusion at the 1% level of significance.

13. $n = 8$, $\overline{X} = 15.7625$, $\sigma = 2$, $SE(\overline{X}) = 0.7071$, $\alpha = 0.05$, $Z_{\alpha/2} = 1.96$.

   H_0: $\mu = 15$ against H_1: $\mu \neq 15$

   Test statistic: $Z = 1.0783$

   p-value $= 2P\{Z > 1.0783\} = 2(1 - 0.8599) = 0.2802$.

The null hypothesis can be rejected at a level of significance
of 0.2802 or higher.

15. (*Initial solution to problem 6*)

$n = 7$, $\overline{X} = 14.7428$, $\sigma = 2$, $SE(\overline{X}) = 0.7559$, $\alpha = 0.05$, $Z_{\alpha/2} = 1.96$.

H_0: $= 14$ against H_1: $\mu \neq 14$

Test statistic: $Z = 0.9826$.

p-value $= 2P\{Z > 0.9826\} = 2(1 - 0.8365) = 0.327$.

(a) When the null hypothesis is rejected, then $|\overline{X} - 14|/SE(\overline{X}) \geq Z_{\alpha/2}$

or equivalently, $|\overline{X} - 14| \geq (1.96)(0.7559) = 1.4816$. Thus,

$\overline{X} \geq 14 + 1.4816 = 15.4816$ or $\overline{X} \leq 14 - 1.4816 = 12.5184$. For us

to reject the null hypothesis when the actual value sent is 15, then

$P\{\text{rejecting } H_0\} = P\{\overline{X} \geq 15.4816\} + P\{\overline{X} \leq 12.5184\}$

$= P\{Z \geq (15.4816 - 15)/0.7559\} + P\{Z \leq (12.5184 - 15)/0.7559\}$

$= P\{Z \geq 0.6371\} + P\{Z \leq -3.283\} = (1 - 0.7389) + (1 - 0.9995)$

$= 0.2616$.

(b) When the null hypothesis is rejected, then $|\overline{X} - 14|/SE(\overline{X}) \geq Z_{\alpha/2}$

or equivalently, $|\overline{X} - 14| \geq (1.96)(0.7559) = 1.4816$. Thus,

$\overline{X} \geq 14 + 1.4816 = 15.4816$ or $\overline{X} \leq 14 - 1.4816 = 12.5184$. For us to reject the null hypothesis when the actual value sent is 13, then

$$P\{\text{rejecting } H_0\} = P\{\overline{X} \geq 15.4816\} + P\{\overline{X} \leq 12.5184\}$$

$$= P\{Z \geq (15.4816 - 13)/0.7559\} + P\{Z \leq (12.5184 - 13)/0.7559\}$$

$$= P\{Z \geq 3.2826\} + P\{Z \leq -0.6371\} = (1 - 0.9995) + (1 - 0.7389)$$

$$= 0.2616.$$

(c) When the null hypothesis is rejected, then $|\overline{X} - 14|/\text{SE}(\overline{X}) \geq Z_{\alpha/2}$

or equivalently, $|\overline{X} - 14| \geq (1.96)(0.7559) = 1.4816$. Thus,

$$\overline{X} \geq 14 + 1.4816 = 15.4816 \text{ or } \overline{X} \leq 14 - 1.4816 = 12.5184. \text{ For us}$$

to reject the null hypothesis when the actual value sent is 16, then

$$P\{\text{rejecting } H_0\} = P\{\overline{X} \geq 15.4816\} + P\{\overline{X} \leq 12.5184\}$$

$$= P\{Z \geq (15.4816 - 16)/0.7559\} + P\{Z \leq (12.5184 - 16)/0.7559\}$$

$$= P\{Z \geq -0.6858\} + P\{Z \leq -4.6059\} = 0.7549 + (1 - 1)$$

$$= 0.7549.$$

SECTION 9.3.1   ONE-SIDED TESTS

PROBLEMS

1. (a) $n = 16$, $\overline{X} = 7.2$, $\sigma = 1.2$, $SE(\overline{X}) = 0.3$, $\alpha = 0.05$, $Z_\alpha = 1.645$.

   H_0: $\mu \geq 7.6$ against H_0: $\mu < 7.6$

   Test statistic: $Z = -1.3333$

   Conclusion: Since $Z = -1.3333$ is not less than -1.645, do not reject H_0. That is, there is insufficient sample evidence to claim that the average weight of the fish is less than 7.6 pounds.

   (b) $n = 16$, $\overline{X} = 7.2$, $\sigma = 1.2$, $SE(\overline{X}) = 0.3$, $\alpha = 0.01$, $Z_\alpha = 2.33$.

   H_0: $\mu \geq 7.6$ against H_0: $\mu < 7.6$

   Test statistic: $Z = -1.3333$

   Conclusion: Since $Z = -1.3333$ is not less than -2.33, do not reject H_0. That is, there is insufficient sample evidence to claim that the average weight of the fish is less than 7.6 pounds.

   (c) p-value = $P\{Z < -1.333\} = (1 - 0.9082) = 0.0918$.

3. (a) $n = 20$, $\overline{X} = 108$, $\sigma = 5$, $SE(\overline{X}) = 1.118$.

Ho: $\mu \leq 100$ against H$_1$: $\mu > 100$

Test statistic: $Z = 7.1554$

p-value $= P\{Z > 7.1554\} = 0$.

(b) $n = 20$, $\overline{X} = 108$, $\sigma = 10$, $SE(\overline{X}) = 2.2361$.

Ho: $\mu \leq 100$ against H$_1$: $\mu > 100$

Test statistic: $Z = 3.5777$

p-value $= P\{Z > 3.5777\} \approx 0$.

(c) $n = 20$, $\overline{X} = 108$, $\sigma = 15$, $SE(\overline{X}) = 3.3541$.

Ho: $\mu \leq 100$ against H$_1$: $\mu > 100$

Test statistic: $Z = 2.3851$

p-value $= P\{Z > 2.3851\} = (1 - 0.9916) = 0.0084$.

5. (a) $n = 2500$, $\overline{X} = 2.95$, $\sigma = 1$, $SE(\overline{X}) = 0.02$, $\alpha = 0.05$, $Z_\alpha = 1.645$.

Ho: $\mu \geq 3$ against : H$_1 < 3$

Test statistic: $Z = -2.5$

Conclusion: Since $Z = -2.25 < -Z_\alpha = -1.645$, Ho is rejected. That is, the data are consistent with the alternative hypothesis that the average number of cavities of children is less than 3 when they use the toothpaste in their cavity-prone years at the 5% level of significance.

(b) Yes.

7. No. Assuming that the sample is random, the p-value (0.305) is large enough to convince one that average nicotine content is not less than 1.5 milligrams per cigarette.

9. $n = 100$, $\overline{X} = 5.6$, $\sigma = 0.14$, $SE(\overline{X}) = 0.014$.

   H_0: $H \geq 6$ against H_1: $\mu < 6$

   Test statistic: $Z = -28.5714$

   p-value = $P\{Z \leq -28.5714\} = 0$

   Conclusion: Since p-value = 0, H_0 is rejected. That is, the data are consistent with the alternative hypothesis that the average amount of soft drink dispensed by the machine is less than 6 ounces.

SECTION 9.4   THE t-TEST FOR THE MEAN OF A NORMAL POPULATION HAVING AN UNKNOWN VARIANCE

PROBLEMS

1. $n = 25$, $\overline{X} = 19.7$, $S = 1.3$, $SE(\overline{X}) = 0.26$, $\alpha = 0.05$, $t_{24, 0.025} = 2.064$,

   H_0: $\mu = 20$ against H_1: $\mu \neq 20$

   Test statistic: $T = -1.1538$

   Conclusion: Since $|T| = 1.1538 < t_{24, 0.025} = 2.064$, H_0 is not rejected at the 5% level of significance. That is, the data are not inconsistent with the null hypothesis that the average amount of phenobarbital in each capsule is equal to 20 milligrams. Thus, the evidence is not strong enough to discredit the manufacturer's claim at the 5% level of significance.

3. (a) $n = 10$, $\overline{X} = 110$, $\sigma = 15$, $SE(\overline{X}) = 4,7434$, $\alpha = 0.05$, $Z_{0.025} = 1.96$,

   H_0: $\mu = 100$ against H_1: $\mu \neq 100$

   Test statistic: $Z = 2.1082$

   Conclusion: Since $|Z| = 2.1082 > Z_{0.025} = 1.96$, H_0 is rejected at the 5% level of significance. That is, the data are consistent with the alternative hypothesis that the mean for the population is not equal to 100.

(b) $n = 10$, $\overline{X} = 110$, $S = 15$, $SE(\overline{X}) = 4.7434$, $\alpha = 0.05$, $t_{9,\,0.025} = 2.262$,

H_0: $\mu = 100$ against H_1: $\mu \neq 100$

Test statistic: $T = 2.1082$

Conclusion: Since $|T| = 2.1082 < t_{9,\,0.025} = 2.262$, H_0 is not rejected at the 5% level of significance. That is, there is insufficient same evidence to conclude that the mean for the population is not equal to 100 at the 5% level of significance.

5. (a) $n = 36$, $\overline{X} = 56.4$, $S = 5.1$, $SE(\overline{X}) = 0.85$, $t_{35,\,0.05} \approx 1.697$.

H_0: $\mu = 55$ against H_1: $\mu \neq 55$

Test statistic: $T = 1.6471$

Conclusion: Since $|T| = 1.6471 < t_{35,\,0.05} = 1.697$, H_0 is not rejected at the 10% level of significance. That is, there is insufficient sample evidence to conclude that the mean depth is not equal to 55 fathoms.

Note: If you use $t_{35,\,0.05} \approx Z = 1.645$, then the null hypothesis will be rejected at the 5% level of significance.

(b) $n = 36$, $\overline{X} = 56.4$, $S = 5.1$, $SE(\overline{X}) = 0.85$, $t_{35,\,0.025} \approx 2.042$.

H_0: $\mu = 55$ against H_1: $\mu \neq 55$

Test statistic: $T = 1.6471$

Conclusion: Since $|T| = 1.6471 < t_{35, 0.025} \approx 2.042$, H_0 is not rejected at the 5% level of significance. That is, there is insufficient sample evidence to conclude that the mean depth is not equal to 55 fathoms.

Note: If you use $t_{35, 0.025} \approx Z = 1.96$, then the null hypothesis will not be rejected at the 5% level of significance.

(c) $n = 36$, $\overline{X} = 56.4$, $S = 5.1$, $SE(\overline{X}) = 0.85$, $t_{35, 0.005} \approx 2.750$.

H_0: $\mu = 55$ against H_1: $\mu \neq 55$

Test statistic: $T = 1.6471$

Conclusion: Since $|T| = 1.6471 < t_{35, 0.005} \approx 2.750$, H_0 is not rejected at the 1% level of significance. That is, there is insufficient sample evidence to conclude that the mean depth is not equal to 55 fathoms.

Note: If you use $t_{35, 0.005} \approx Z = 2.575$, then the null hypothesis will not be rejected at the 1% level of significance.

7. $n = 28$, $\overline{X} = 1.0$, $S = 0.3$, $SE(\overline{X}) = 0.0567$, $\alpha = 0.05$, $t_{27, 0.025} = 2.052$.

H_0: $\mu = 0.8$ against H_1: $\mu \neq 0.8$

Test statistic: $T = 3.5273$

Conclusion: Since $[T] = 3.5273 > t_{27, 0.025} = 2.052$, H_0 is rejected at the 5% level of significance. That is, there is sufficient sample evidence to conclude that the mean response time is not equal to 0.8 seconds. Thus we can conclude that alcohol consumption does affect the response time to the stimulus for that particular species of pigs.

9. Results **will** vary.

11. $n = 20$, $\overline{X} = 22.8$, $S = 1.4$, $SE(\overline{X}) = 0.3130$, $\alpha = 0.05$, $t_{19,\,0.05} = 1.729$.

H_0: $\mu \geq 23$ against H_1: $\mu < 23$

Test statistic: $T = -0.6390$

Conclusion: Since $T = -0.6390 > -t_{19,\,0.05} = -1.729$, H_0 is not rejected at the 5% level of significance. That is, the data are not inconsistent with the null hypothesis that the average weight of the loaves of bread is greater than or equal to 23 ounces. The judge should rule for the bakery.

13. $n = 8$, $\overline{X} = 29.25$, $S = 2.9641$, $SE(\overline{X}) = 1.0480$, $t_{7,\,0.05} = 1.895$,

(a) H_0: $\mu \geq 31$

(b) H_1: $\mu < 31$

(c) Test statistic: $T = -1.6698$.

Conclusion: Since $T = -1.6698 > -t_{7,\,0.05} = -1.895$, H_0 is not rejected at the 5% level of significance. That is, the data are not inconsistent with the null hypothesis that the average miles per gallon on the highway to be at least 31.

(d) Conclusion: Since $T = -1.6698 > -t7, 0.01 = -2.998$, H_0 is not rejected at the 1% level of significance. That is, the data are not inconsistent with the null hypothesis that the average miles per gallon on the highway to be at least 31.

15. $n = 25$, $\overline{X} = 367$, $S = 62$, $SE(\overline{X}) = 12.4$.

   H_0: $\mu \geq 400$ against H_1: $\mu < 400$

   Test statistic: $T = -2.6613$

   p-value $= P\{|T| > 2.6613\} \Rightarrow 0.005 \leq$ p-value ≤ 0.01

   Conclusion: For significance levels (approximately) greater than or equal 0.01, the null hypothesis will be rejected.

   Note: The value of 0.01 is an approximate lower bound for the significance level.

17. $n = 100$, $\overline{X} = 98.74$, $S = 1.1$, $SE(\overline{X}) - 0.11$, $t_{99,\,0.05} \approx Z_{0.05} = 1.645$, $t_{99,\,0.05} \approx Z_{0.01} = 2.33$.

   (a) H_0: $\mu \leq 98.6$ against H_1: $\mu > 98.60$

   Test statistic: $T = 1.2727$

   Conclusion: Since $T = 1.2727 < Z_{0.05} = 1.645$, H_0 is not rejected at the 5% level of significance. That is, there is insufficient sample evidence to conclude that the average body temperature is now greater than 98.6 °F.

   (b) Conclusion: Since $T = 1.2727 < Z_{0.05} = 2.33$, H_0 is not rejected at the 1% level of significance. That is, there is insufficient sample evidence to conclude that the average body temperature is now greater than 98.6 °F.

SECTION 9.5   HYPOTHESIS TESTS CONCERNING POPULATION PROPORTIONS

PROBLEMS

1. $n = 50$, $x = 42$, $p_0 = 0.72$, $\mu = np_0 = 36$, $\sigma = \sqrt{np_0(1-p_0)} = 3.1749$.

H_0: $p \leq 0.72$ against H_1: $p > 0.72$

Test statistic: $X = 42$

p-value $= P\{X \geq 42\} = P\{X \geq 41.5\}$

$$\approx P\{Z \geq (41.5 - 36)/3.1749\} = P\{Z \geq 1.7323\}$$

$$= 1 - 0.9582 = 0.0418.$$

3. $n = 300$, $x = 18$, $p_0 = 1/14 = 0.0714$, $\mu = np_0 = 21.4286$,

$\sigma = \sqrt{np_0(1-p_0)} = 4.4608$, $\alpha = 0.05$.

H_0: $p \geq 0.0714$ against H_1: $p < 0.0714$

Test statistic: $X = 18$

p-value $= P\{X \leq 18\} = P\{X \leq 18.5\}$

$$\approx P\{Z \leq (18.5 - 21.4286)/4.4608\}$$

$$= P\{Z \leq -0.6565\} = 1 - 0.7486 = 0.2514$$

Conclusion: Since p-value $=0.2514 > \alpha = 0.05$, H_0 is not rejected at the 5% level of significance. That is, there is insufficient sample data to conclude that the proportion of shoplifters is less than 0.0714. Apparently, the policy is not working.

5. (a) H_0: $p \leq 0.5$ against H_1: $p > 0.5$

   (b) $n = 100$, $x = 56$, $p_0 = 0.5$, $\mu = np_0 = 50$, $\sigma = \sqrt{np_0(1-p_0)} = 5$.

   Test statistic: $X = 56$

   p-value $= P\{X \geq 56\} = P\{X \geq 55.5\}$

   $\approx P\{Z \geq (55.5 - 50)/5\} = P\{Z \geq 1.1\} = 1 - 0.8643 = 0.1357$.

   (c) $n = 200$, $x = 112$, $p_0 = 0.5$, $\mu = np_0 = 100$, $\sigma = \sqrt{np_0(1-p_0)} = 7.0711$.

   Test statistic : $X = 112$

   p-value $= P\{X \geq 112\} = P\{X \geq 111.5\}$

   $\approx P\{Z \geq (111.5 - 100)/7.0711\} = P\{Z \geq 1.6263\}$

   $= 1 - 0.9484 = 0.0516$.

   (d) $n = 500$, $x = 280$, $p_0 = 0.5$, $\mu = np_0 = 250$,

   $\sigma = \sqrt{np_0(1-p_0)} = 11.1803$.

   Test statistic: $X = 280$

   p-value $= P\{X \geq 280\} = P\{X \geq 279.5\} \approx P\{Z \geq (279.5 - 250)/11.1803\}$

   $= P\{Z > 2.6386\} = 1 - 0.9959 = 0.0041$.

As n increases the p-value decreases, because we have more confidence in the sample data for larger sample size n.

7. (a) $n = 100$, $x = 56$, $p_0 = 0.5$, $\mu = np_0 = 50$, $\sigma = \sqrt{np_0(1-p_0)} = 5$,

   $\alpha = 0.05$.

   $H_0: p \le 0.5$ against $H_1: p > 0.5$

   Test statistic: $X = 56$

   p-value $= P\{X \ge 56\} = P\{X \ge 55.5\}$

   $\approx P\{Z \ge (55.5\text{-}50)/5\} = P\{Z \ge 1.1\}$

   $= 1 - 0.8643 = 0.1357$.

   Conclusion: Since p-value $= 0.1357 > \alpha = 0.05$, H_0 is not rejected at the 5% level of significance. That is, there is insufficient sample evidence to conclude that more than 50% of the population supports the initiative concerning limitations on driving automobiles in the downtown area at the 5% level of significance.

   (b) $n = 120$, $x = 68$, $p_0 = 0.5$, $\mu = np_0 = 60$, $\sigma = \sqrt{np_0(1-p_0)} = 5.4772$,

   $\alpha = 0.05$.

   $H_0: p \le 0.5$ against $H_1: p > 0.5$

   Test statistic: $X = 68$

   p-value $= P\{X \ge 68\} = P\{X \ge 67.5\}$

   $\approx P\{Z \ge (67.5 - 60)/5.4772\} = P\{Z \ge 1.3693\}$

$$= 1 - 0.9147 = 0.0853.$$

Conclusion: Since p-value = 0.0853 > α = 0.05, H_0 is not rejected at the 5% level of significance. That is, there is insufficient sample evidence to conclude that more than 50% of the population supports the initiative concerning limitations on driving automobiles in the downtown area at the 5% level of significance.

(c)   n = 110, x = 62, p_0 = 0.5, μ = np_0 = 55, $\sigma = \sqrt{np_0(1-p_0)}$ = 5.2440,

α = 0.05.

H_0: p $\le$ 0.5 against H: p > 0.5

Test statistic: X = 62

p-value = $P\{X \ge 62\}$ = $P\{X \ge 61.5\}$

$\approx P\{Z \ge (61.5 - 55)/5.244\}$ = $P\{Z \ge 1.2395\}$

$= 1 - 0.8925 = 0.1075.$

Conclusion: Since p-value = 0.1075 > α = 0.05, H_0 is not rejected at the 5% level of significance. That is, there is insufficient sample evidence to conclude that more than 50% of the population supports the initiative concerning limitations on driving automobiles in the downtown area at the 5% level of significance.

(d) $n = 330$, $x = 186$, $p_0 = 0.5$, $\mu = np_0 = 165$, $\sigma = \sqrt{np_0(1-p_0)} = 9.0830$,

$\alpha = 0.05$.

H_0: $p \le 0.5$ against H_1: $p > 0.5$

Test statistic: $X = 56$

p-value $= P\{X \ge 186\} = P\{X \ge 185.5\}$

$\approx P\{Z \ge (185.5 - 165)/9.083\} = P\{Z \ge 2.257\}$

$= 1 - 0.9881 = 0.0119$.

Conclusion: Since p-value $= 0.0119 < \alpha = 0.05$, H_0 is rejected at the 5% level of significance. That is, there is sufficient sample evidence to conclude that more than 50% of the population supports the initiative concerning limitations on driving automobiles in the downtown area at the 5% level of significance.

9. $n = 90$, $x = 5$, $p_0 = 0.04$, $\mu = np_0 = 3.6$, $\sigma = \sqrt{np_0(1-p_0)} = 1.859$,

$\alpha = 0.05$. *Note:* $np_0 = 3.6 < 5$. Normality assumption is not met.

H_0: $p \le 0.04$ against H_1: $p > 0.04$

Test statistic: $X = 5$

p-value $= P\{X \ge 5\} = P\{X \ge 4.5\}$

$\approx P\{Z \ge (4.5 - 3.6)/1.859\} = P\{Z > 0.4841\}$

$= 1 - 0.6844 = 0.3156$.

Conclusion: Since p-value = 0.3156 > α = 0.05, H_0 is not rejected at the 10% level of significance. That is, there is insufficient sample evidence to conclude that the proportion of defectives is more than 4%. Same conclusion at the 5% level of significance.

11. n = 264, x = 65, p_0 = 0.22, $\mu = np_0$ = 58.08, $\sigma = \sqrt{np_0(1-p_0)}$ = 6.7307,

α = 0.05.

H_0: p $\le$ 0.22 against H_1: p > 0.22

Test statistic: X = 65

p-value $\doteq$ P{X $\ge$ 65} = P{X $\ge$ 64.5}

$\approx$ P{Z $\ge$ (64.5 - 58.08)/6.7307} = P{Z $\ge$ 0.9538}

= 1 - 0.8289 = 0.1711.

Conclusion: Since p-value =0.1711 > α = 0.05, H_0 is not rejected at the 5% level of significance. That is, the data are not inconsistent with the null hypothesis that the proportion of students at Berkeley who classify themselves as liberals to be less than or equal to 22%.

13. (a) n = 30, x = 19, p_0 = 0.5, $\mu = np_0$ = 15, $\sigma = \sqrt{np_0(1-p_0)}$ = 2.7386.

α = 0.1.

H_0: p = 0.5 against H_1: p $\ne$ 0.5

Test statistic: X = 19

p-value = 2[Min(P{X $\le$ 19}, P{X $\ge$ 19}

Now, $P\{X \le 19\} = P\{X \le 19.5\} \approx P\{Z \le (19.5 - 15)/2.7386\}$

$= P\{Z \le 1.6432\} = 0.9495$, and $P\{X \ge 54.5\} \approx 0.0505$. Hence,

p-value = 2(0.0505) = 0.1010.

Conclusion: Since p-value = 0.1010 > α = 0.1, H_0 is not rejected at the 10% level of significance. That is, the data are inconsistent with the alternative hypothesis that the proportion of the time that the person is "lucky" is not equal to 50%.

(b) When α = 0.05, we do not reject the null hypothesis.

(c) p-value = 0.1010.

15. n = 120, x = 48, p_0 = 0.25, μ = np_0 = 30,

$$\sigma = \sqrt{np_0(1-p_0)} = 4.7434.$$

H_0: p = 0.25 against H_1: p $\ne$ 0.25

Test statistic: X = 48

p-value = 2[Min(P$\{X \le 48\}$, P$\{X \ge 48\}$

Now, $P\{X \le 48\} = P\{X \le 48.5\} \approx P\{Z \le (48.5 - 30)/4.7434\}$

$= P\{Z \le 3.9001\} = 0.9999$, and $P\{X \ge 48\} = 0.0001$. Hence,

p-value = 2(0.0001) = 0.0002.

REVIEW PROBLEMS

1. (b); statement (a) ignores whether the evidence of the data was

   significant.

3. (a) No. The significance level or p-value of the test is not specified.

   (b) We could interpret the statement to be true at the 5% level of significance since most reports in the news use this level of significance without stating it.

5. (a) $n = 10$, $\overline{X} = 38.17$, $S = 2.9721$, $SE(\overline{X}) = 0.9399$, $t_{9, 0.01} = 2.821$.

   H_0: $\mu \geq 40$ against H_1: $\mu < 40$

   Test statistic: $T = 1.9470$

   Conclusion: Since $T = -1.947 > -t_{9, 0.01} = -2.821$, H_0 is not rejected at the 1% level of significance. That is, there is insufficient sample evidence to conclude that the average time for the new route is shorter.

   (b) $t_{9, 0.05} = 1.833$.

   Conclusion: Since $T = -1.947 < -t_{9, 0.05} = -1.833$, H_0 is rejected at the 5% level of significance. That is, there is sufficient sample evidence to conclude that the average time for the new route is shorter at the 5% level of significance.

   (c) $t_{9, 0.1} = 1.383$.

   Conclusion: Since $T = -1.947 < -t_{9, 0.1} = -1.383$, H_0 is rejected at the 5% level of significance. That is, there is sufficient sample evidence to conclude that the average time for the new route is shorter at the 5% level of significance.

7. $n = 20$, $\overline{X} = 198.4$, $S = 11.3712$, $SE(\overline{X}) = 2.5427$, $t_{19, 0.05} = 1.729$.

   H_0: $\mu \geq 200$ against H1: $\mu < 200$

   Test statistic: $T = -0.6293$

   Conclusion: Since $T = -0.6293 > - t_{19, 0.05} = -1.729$, H_0 is not rejected at the 5% level of significance. That is, there is insufficient sample evidence to conclude that the average blood cholesterol level for the entering college students is less than 200.

9. One would probably rule against Mr. Caputo. If the drawing is done at random, one would expect a 50% chance of either a Democrat's or a Republican's name being selected. Thus, if a two-tail test for a single proportion was done where the test proportion is 50% (0.5), the p-value for the test will be approximately zero. This would indicate that the true proportion is very significantly different from 50% which would indicate, maybe, that the process was not fair.

11. Answers will vary.

13. The natural logarithms for the closing prices for the seventeen closing prices and the successive differences of these logarithms are shown below.

CLOSING PRICES, X	LN(X)	DIFFERENCES
387.10	5.9587	-0.0023
388.00	5.9610	-0.0042
389.65	5.9652	-0.0035
391.00	5.9687	-0.0024
391.95	5.9711	0.0023
391.05	5.9688	0.0039
389.50	5.9649	-0.0019
390.25	5.9668	0.0058
388.00	5.9610	-0.0077
391.00	5.9687	0.0013
390.50	5.9674	-0.0090
394.00	5.9764	-0.0025
395.00	5.9789	0.0038
393.50	5.9751	-0.0069
396.25	5.9820	0.0000
396.25	5.9820	0.0020
395.45	5.9800	

$n = 16$, $\overline{X} = -0.0013$, $S = 0.0044$, $SE(\overline{X}) = 0.0011$, $t_{15, 0.025} = 2.131$ when

the significance level $a = 0.05$ (since none was specified).

H_0: $\mu = 0$ against H_1: $\mu \neq 0$

Test statistic: $T = -1.1818$

Conclusion: Since $|T| = 1.1818 < t_{15, 0.025} = 2.131$, H_0 is not rejected at the 5% level of significance. That is, there is insufficient sample evidence to conclude that the mean daily average of the successive logarithm differences of the closing price of the stock is not equal to zero.

Chapter 10 HYPOTHESIS TESTS CONCERNING TWO POPULATIONS

SECTION 10.2 TESTING EQUALITY OF MEANS OF TWO NORMAL POPULATIONS: CASE OF KNOWN VARIANCES

PROBLEMS

1. (a) $n = 12$, $m = 14$, $\overline{X} = 45.2$, $\overline{Y} = 48.6$, $\sigma_x = 0.8$, $\sigma_y = 1.0$,

$$\sqrt{\frac{\sigma_x^2}{n} + \frac{\sigma_y^2}{m}} = 0.3532, \; \alpha = 0.05, \; Z_{\alpha/2} = 1.96.$$

H_0: $\mu_x = \mu_y$ against H_1: $\mu_x \neq \mu_y$

Test statistic: $Z = (45.2 - 48.6)/(0.3532) = -9.626$

Conclusion: Since $|Z| = 9.626 > Z_{\alpha/2} = 1.96$, reject H_0. That is, the mean yield for the two methods is not the same at the 5% level of significance.

(b) p-value = $2P\{Z > 9.626\} = 0.0$.

3. (a) $n = m = 10$, $\overline{X} = 122.541$, $\overline{Y} = 122.494$, $\sigma_x = \sigma_y = 0.5$,

$$\sqrt{\frac{\sigma_x^2}{n} + \frac{\sigma_y^2}{m}} = 0.2236, \; \alpha = 0.05, \; Z_{\alpha/2} = 1.96.$$

$H_0: \mu_x = \mu_y$ against $H_1: \mu_x \neq \mu_y$

Test statistic: $Z = (122.541 - 122.494)/(0.2236) = 0.2102$

Conclusion: Since $|Z| = 0.2102 < Z_{\alpha/2} = 1.96$, do not reject H_0. That is, there is insufficient sample evidence to support the claim that mean lengths of the cuttings are not the same at the 5% level of significance.

(b) p-value $= 2P\{Z > 0.2102\} = 2(1 - 0.5832) = 0.8336$.

5. You can interchange the (arbitrary) subscripts for the data sets and use the given hypothesis test.

7. $n = m = 9$, $\overline{X} = 5.6$, $\overline{Y} = 4.1$, $\sigma_x = \sigma_y = 2$,

$$\sqrt{\frac{\sigma_x^2}{n} + \frac{\sigma_y^2}{m}} = 0.9428, \; \alpha = 0.01, \; Z_{\alpha} = 2.33.$$

$H_0: \mu_x \leq \mu_y$ against $H_1: \mu_x > \mu_y$

Test statistic: $Z = (5.6 - 4.1)/(0.9428) = 1.5910$

Conclusion: Since $Z = 1.591 < Z_\alpha = 2.33$, do not reject H_0. That is, there is insufficient sample evidence to support the claim that mean length of the first message will be greater than the mean length of the second message at the 1% level of significance.

SECTION 10.3 TESTING EQUALITY OF MEANS:
UNKNOWN VARIANCES AND LARGE
SAMPLE SIZES

PROBLEMS

1. $n = m = 35$, $\overline{X} = 72.6$, $\overline{Y} = 74.0$, $S_x^2 = 6.6$, $S_y^2 = 6.2$,

$$\sqrt{\frac{S_x^2}{n} + \frac{S_y^2}{m}} = 0.6047, \; \alpha = 0.05, \; Z_{\alpha/2} = 1.96.$$

H_0: $\mu_x = \mu_y$ against H_1: $\mu_x \neq \mu_y$

Test statistic: $Z = (72.6 - 74)/(0.6047) = -2.3152$

Conclusion: Since $|Z| = 2.3152 > Z_{\alpha/2} = 1.96$, reject H_0. That
is, there is sufficient sample evidence to support the claim that mean
scores for the two groups of students are not the same at the 5% level of
significance.

p-value = $2P\{Z > 2.3152\} = 2(1 - 0.9898) = 0.0204$.

3. $n = m = 30$, $\overline{X} = 3.6$, $\overline{Y} = 3.8$, $S_x^2 = (1.3)^2$, $S_y^2 = (1.4)^2$,

$$\sqrt{\frac{S_x^2}{n} + \frac{S_y^2}{m}} = 0.3488.$$

H_0: $\mu_x = \mu_y$ against H_1: $\mu_x \neq \mu_y$

Test statistic: $Z = (3.6 - 3.8)/(0.3488) = -0.5734$.

p-value $= 2P\{Z > 0.5734\} = 2(1 - 0.7157) = 0.5686$.

Conclusion: Since p-value $= 0.5686$ (large), do not reject H_0. That is, there is insufficient sample evidence to conclude that the average visits for the two groups of women are significantly different.

5. $n = m = 36$, $\overline{X} = 12.4$, $\overline{Y} = 14.2$, $S_x^2 = (1.6)^2$, $S_y^2 = (1.8)^2$,

$$\sqrt{\frac{S_x^2}{n} + \frac{S_y^2}{m}} = 0.4014.$$

$H_0: \mu_x = \mu_y$ against $H_1: \mu_x \neq \mu_y$

Test statistic: $Z = (12.4 - 14.2)/(0.4014) = -4.4843$.

p-value $= 2P\{Z > 4.4843\} = 2(1 - 1) = 0.0$.

Conclusion: Since p-value $= 0$ (very small), reject H_0. That is, there is sufficient sample information to conclude that the mean yield for the two varieties are significantly different.

7. $n = 100$, $m = 60$, $\overline{X} = 102.2$, $\overline{Y} = 105.3$, $S_x^2 = (11.8)^2$, $S_y^2 = (10.6)^2$,

$$\sqrt{\frac{S_x^2}{n} + \frac{S_y^2}{m}} = 1.8069, \; \alpha = 0.01, \; Z_{\alpha/2} = 2.33.$$

H_0: $\mu_y \leq \mu_x$ against H_1: $\mu_y > \mu_x$

Test statistic: $Z = (105.3 - 102.2)/(1.8069) = 1.7156$.

Conclusion: Since $Z = 1.7156 < Z_\alpha = 2.33$, do not reject H_0. That is, there is insufficient sample evidence to conclude that the average IQ scores for the rural students in upper Michigan is greater than the average scores for the urban students from upper Michigan.

9. $n = 28$, $m = 32$, $\overline{X} = 97.07$ for A, $\overline{Y} = 101.97$ for B, $S_x^2 = (13.62)^2$,

$S_y^2 = (13.74)^2$, $\sqrt{\dfrac{S_x^2}{n} + \dfrac{S_y^2}{m}} = 3.5390$, $\alpha = 0.05$, $Z_\alpha = 1.645$.

H_0: $\mu_B \leq \mu_A$ against H_1: $\mu_B > \mu_A$

Test statistic: $Z = (101.97 - 97.07)/(3.539) = 1.3846$.

Conclusion: Since $Z = 1.3846 < Z_\alpha = 1.645$, do not reject H_0. That is, there is insufficient sample evidence to conclude that the average lifetime of the bulbs from manufacturer B is greater than the average lifetime of the bulbs from manufacturer A.

11. $n = 55$, $m = 72$, $\overline{X} = 10.8$ (female), $\overline{Y} = 12.2$ (male), $S_x^2 = 0.9$,

$S_y^2 = 1.1$, $\sqrt{\dfrac{S_x^2}{n} + \dfrac{S_y^2}{m}} = 0.1779$.

(a) H_0: $\mu_f \geq \mu_m$ against H_1: $\mu_f < \mu_m$

(b) $Z = (10.8 - 12.2)/0.1779 = -7.8696$.

p-value $= P\{Z > 7.8696\} = 0.0$.

(c) Conclusion: Since p-value $= 0.0$ (very small), reject H_0. That is, there is sufficient sample evidence to conclude that the average salary for a female is less than the average salary for a male at any level of significance.

13. (a) $n = m = 20$, $\overline{X} = 70.6$, $\overline{Y} = 77.4$, $S_x^2 = (8.4)^2$,

$S_y^2 = (7.4)^2$, $\sqrt{\dfrac{S_x^2}{n} + \dfrac{S_y^2}{m}} = 2.5032$, $\alpha = 0.01$, $Z_\alpha = 2.58$.

H_0: $\mu_x = \mu_y$ against H_1: $\mu_x \neq \mu_y$

Test statistic: $Z = (70.6 - 77.4)/(2.5032) = -2.7165$.

Conclusion: Since $Z = -2.7165 < -Z_\alpha = -2.58$, reject H_0. That is, there is sufficient sample evidence to conclude that the average time for the supervised group is different than that for the unsupervised group at the 1% level of significance.

(b) p-value $= P\{Z > 2.7165\} = 1 - 0.9967 = 0.0033$.

(c) Supervision seemed to have helped to reduce the average score.

SECTION 10.4 TESTING EQUALITY OF MEANS: SMALL SAMPLE TESTS WHEN THE UNKNOWN POPULA TION VARIANCES ARE EQUAL

PROBLEMS

1. (a) $n = 11$, $m = 14$, $\overline{X} = 129.18$, $\overline{Y} = 123.36$, $S_x^2 = (5.72)^2$, $S_y^2 = (5.73)^2$,

    $\alpha = 0.01$, $t_{23,\,\alpha/2} = 2.807$, $S_p^2 = (5.73)^2$.

    H_0: $\mu_x = \mu_y$ against H_1: $\mu_x \neq \mu_y$

    Test statistic: $T = (129.18 - 123.36)/(2.3087) = 2.5209$.

    Conclusion: Since $|T| = 2.5209 < 2.807$, do not reject H_0. That is, the data are inconsistent with the alternative hypothesis that the average systolic blood pressure for smokers and nonsmokers are significantly different at the 1% level of significance.

    (b) $\alpha = 0.05$, $t_{23,\,\alpha/2} = 2.069$

    Same conclusion as in part (a).

3. $n = m = 9$, $\overline{X} = 14.467$, $\overline{Y} = 14.611$, $S_x^2 = (0.534)^2$, $S_y^2 = (0.42)^2$,

    $\alpha = 0.05$, $t_{16,\,\alpha/2} = 2.120$, $S_p^2 = (0.480)^2$.

    H_0: $\mu_x \geq \mu_y$ against H_1: $\mu_x < \mu_y$

    Test statistic: $T = (14.467 - 14.611)/(38.2264) = -0.6364$.

Conclusion: Since T = -0.6364 < 2.120, reject H_0. That is, the data are consistent with the alternative hypothesis that the average dissolve time for the "generic" brand is less than the average dissolve time for the "name" brand.

(b) $\alpha = 0.1$, $t_{16, \alpha/2} = 1.714$.

   Same conclusion as in part (a).

5. $n = 10$, $m = 10$, $\overline{X} = 65.6$, $\overline{Y} = 70.4$, $S_x^2 = (5.4)^2$, $S_y^2 = (4.8)^2$,

   $\alpha = 0.05$, $t_{18, \alpha} = 1.734$, $S_p^2 = 26.1$.

   H_0: $\mu_y \leq \mu_x$ against H_1: $\mu_y > \mu_x$

   Test statistic: T = (70.4 - 65.6)/(2.2847) = 2.1009.

   Conclusion: Since T = 2.1009 > 1.734, reject H_0. That is, the data are consistent with the alternative hypothesis that the average score for the students using the experimental method is greater than the average score for the students using the standard method.

7. $n = 6$, $m = 14$, $\overline{X} = 26.8$, $\overline{Y} = 42.5$, $S_x^2 = (9.2)^2$, $S_y^2 = (9.5)^2$,

   $\alpha = 0.05$, $t_{18, \alpha/2} = 2.101$, $S_p^2 = 88.6917$.

   H_0: $\mu_x = \mu_y$ against H_1: $\mu_x \neq \mu_y$

   Test statistic: T = (26.8 - 42.5)/(4.5953) = -3.4165.

Conclusion: Since T| = 3.4165 > 2.101, reject H_0. That is, the data are consistent with the alternative hypothesis that the average dietary fiber consumed daily is not the same for the population of vegetarians having diverticular disease and the population of vegetarians who do not have this disease.

9. (a) $n = 10$, $m = 9$, $\overline{X} = 11.77$, $\overline{Y} = 13.74$, $S_x^2 = (1.61)^2$, $S_y^2 = (1.45)^2$,

   $\alpha = 0.1$, $t_{17, \alpha/2} = 1.74$, $S_p^2 = (1.54)^2$.

   $H_0: \mu_x = \mu_y$ against $H_1: \mu_x \neq \mu_y$

   Test statistic: $T = (11.77 - 13.74)/(0.7076) = -2.7841$.

   Conclusion: Since |T[= 2.7841 > 1.74, reject H_0. That is, the data are consistent with the alternative hypothesis that the means for the two populations are not equal at the 5% level of significance.

(b) $\alpha = 0.05$, $t_{17, \alpha/2} = 2.11$. Same conclusion as in (a).

(c) $\alpha = 0.01$, $t_{17, \alpha/2} = 2.898$. Do not reject the null hypothesis. There is insufficient sample evidence to conclude that the means are different at the 1% level of significance.

SECTION 10.5 PAIRED-SAMPLE *t* TEST

PROBLEMS

1. (a) Difference used: Heart rate before use - Heart rate after use.

    $n = 12$, $\overline{D} = -3.75$, $S_d = 3.079$, $\alpha = 0.05$, $t_{11, \alpha/2} = 2.201$.

    H_0: $\mu_d = 0$ against H_1: $\mu_d \neq 0$

    Test statistic: $T = -4.22$.

    Conclusion: Since $|T| = 4.22 > 2.201$, reject H_0. That is, the data are consistent with the alternative hypothesis that chewing smokeless tobacco does result in a change in the mean heart rate of the population of regular users of chewing tobacco.

    (b) p-value - $2P\{T > 4.22\} \approx 2(0.001) = 0.002$ with df = 11.

3. Difference used: X_i - Y_i.

    $n = 11$, $\overline{D} = 0.09$, $S_d = 8.18$, $\alpha = 0.05$, $t_{10, \alpha/2} = 2.228$.

    H_0: $\mu_d = 0$ against H_1: $\mu_d \neq 0$

    Test statistic: $T = 0.04$.

    Conclusion: Since $|T[== 0.04 < 2.228$, do not reject H_0. That is, the data are not inconsistent with the null hypothesis that the average difference is equal to zero at the 5% level of significance.

    p-value = $2P\{T > 0.04\} > 2(0.4) = 0.8$, with df = 10.

5. Difference used: (IQ test score for twin raised by mother or father)
 - (IQ test score for twin raised by neither parent).

 $n = 14$, $\overline{D} = 1.07$, $S_d = 5.77$, $\alpha = 0.05$, $t_{13, \alpha/2} = 2.160$.

 (a) H_0: $\mu_d = 0$ against H_1: $\mu_d \neq 0$

   Test statistic: $T = 0.69$.

 (b) Conclusion: Since $|T| = 0.69 < 2.16$, do not reject H_0. That is, the data are consistent with the null hypothesis that the average IQ test score of a twin is not affected by whether he or she is raised by a biological parent at the 5% level of significance.

7. Difference used: IQ score before course - IQ score after course.

 $n = 12$, $\overline{D} = -4.92$, $S_d = 5.76$, $\alpha = 0.05$, $t_{11, \alpha/2} = 2.201$.

 H_0: $\mu_d = 0$ against H_1: $\mu_d \neq 0$

 Test statistic: $T = -2.96$.

 Conclusion: Since $|T| = 2.96 > 2.201$, reject H_0. That is, the data are consistent with the alternative hypothesis that the average IQ score will be different after the student takes a statistics course at the 5% level of significance.

9. Difference used: Weight before - Weight after.

 $n = 9$, $\overline{D} = 1.78$, $S_d = 9.87$, $\alpha = 0.05$, $t_{8, \alpha} = 1.86$.

 (a) H_0: $\mu_d \leq 0$ against H_1: $\mu_d > 0$

Test statistic: $T = 0.54$.

(b) Conclusion: Since $T = 0.54 < 1.86$, do not reject H_0. That is, the data are consistent with the null hypothesis that the average weight loss is less than or equal to zero at the 5% level of significance. There is insufficient sample evidence to conclude that the diet program is working.

11. Difference used: 1999 Rate -1990 Rate.

$n = 23$, $\overline{D} = -1.113$, $S_d = 1.0935$, $\alpha = 0.05$, $t_{22, \alpha} = 1.717$.

H_0: $\mu_d \leq 0$ against H_1: $\mu_d > 0$

Test statistic: $T = -4.88$.

p-value $= P\{T > -4.88\}$ with df $= 22$. That is, p-value ≈ 1.

Conclusion: Since the p-value ≈ 1 (large), do not reject H_0. That is, the data are inconsistent with the alternative hypothesis that the average worldwide birth rates in 1999 is greater than those in 1990.

SECTION 10.6 TESTING EQUALITY OF POPULATION PROPORTIONS

PROBLEMS

1. (a) $n_1 = 100$, $n_2 = 100$, $x_1 = 20$, $x_2 = 12$, $\hat{p}_1 = 0.20$, $\hat{p}_2 = 0.12$,

$\hat{p} = 0.16$, $\alpha = 0.05$, $Z_{\alpha/2} = 1.96$.

H_0: $p_1 = p_2$ against H_1: $p_1 \neq p_2$.

Test statistic: $Z = 1.5430$.

Conclusion: Since $|Z| = 1.5430 < 1.96$, do not reject H_0. That is, the data are not inconsistent with the null hypothesis that the proportion of unacceptable transistors for the two methods are the same at the 5% level of significance.

(b) $\alpha = 0.1$, $Z_{\alpha/2} = 1.645$. Same conclusion as in part (a).

3. (a) $n_1 = 300$, $n_2 = 300$, $x_1 = 57$, $x_2 = 36$, $\hat{p}_1 = 0.19$, $\hat{p}_2 = 0.12$,

$\hat{p} = 0.155$, $\alpha = 0.1$, $Z_{\alpha/2} = 1.645$.

H_0: $p_1 = p_2$ against H_1: $p_1 \neq p_2$.

Test statistic: $Z = 2.3689$.

Conclusion: Since $|Z| = 2.3689 > 1.645$, reject H_0. That is, the data are consistent with the alternative hypothesis that the proportion of accidents for the single male and married male policyholders (ages 25-30) are different at the 5% level of significance.

(b) p-value = $2P\{Z > 2.3689\} = 2(1 - 0.9911) = 0.0178$.

5. (a) $n_1 = 100$, $n_2 = 100$, $x_1 = 64$, $x_2 = 52$, $\hat{p}_1 = 0.64$, $\hat{p}_2 = 0.52$,

$\hat{p} = 0.58$, $\alpha = 0.05$, $Z_{\alpha/2} = 1.96$.

H_0: $p_1 = p_2$ against H_1: $p_1 \neq p_2$.

Test statistic: $Z = 1.7192$.

Conclusion: Since $|Z| = 1.7192 < 1.96$, do not reject H_0. That is, there is insufficient sample evidence to conclude that the proportion of insects killed by spray 1 is not equal to the proportion of insects killed by spray 2 at the 5% level of significance.

(b) p-value = $2P\{Z > 1.7192\} = 2(1 - 0.9573) = 0.0854$.

7. $n_1 = 1000$, $n_2 = 1000$, $x_1 = 212$, $x_2 = 272$, $\hat{p}_1 = 0.212$, $\hat{p}_2 = 0.272$,

$\hat{p} = 0.242$, $\alpha = 0.05$, $Z_{\alpha/2} = 1.96$.

H_0: $p_1 = p_2$ against H_1: $p_1 \neq p_2$.

Test statistic: $Z = -3.1325$.

Conclusion: Since $|Z| = 3.1325 > 1.96$, reject H_0. That is, the data are consistent with the alternative hypothesis that the proportion of female scientists are different for the years 1983 and 1990 at the 5% level of significance.

p-value = 2(1 - 0.9991) = 0.0018.

9. $n_1 = 671,976$, $n_2 = 3,046,162$, $x_1 = 330,535$, $x_2 = 1,483,487$,

$\hat{p}_1 = 0.4919$, $\hat{p}_2 = 0.4870$, $\hat{p} = 0.4879$, $\alpha = 0.05$, $Z_{\alpha/2} = 1.96$.

H_0: $p_1 = p_2$ against H_1: $p_1 \neq p_2$.

Test statistic: $Z = 7.2863$.

Conclusion: Since $|Z| = 7.2863 > 1.96$, reject H_0. That is, the data are consistent with the alternative hypothesis that the proportion of African American female babies are not equal to the proportion of white female babies at the 5% level of significance.

p-value = $2P\{Z > 7.2863\} = 0$.

11. (a) $n_1 = 82$, $n_2 = 120$, $x_1 = 78$, $x_2 = 90$, $\hat{p}_1 = 0.9512$, $\hat{p}_2 = 0.75$,

$\hat{p} = 0.8317$, $\alpha = 0.05$, $Z_\alpha = 1.645$.

H_0: $p_1 \leq p_2$ against H_1: $p_1 > p_2$.

Test statistic: $Z = 3.7534$.

Conclusion: Since $Z = 3.7534 > 1.645$, reject H_0. That is, the data are consistent with the alternative hypothesis that the proportion of the population who heard the lecture and use a car seat is more than the proportion who did not hear the lecture at the 5% level of significance.

(b) p-value $= P\{Z \geq 3.7534\} = 0.0$.

13. $n_1 = 102$, $n_2 = 102$, $x_1 = 29$, $x_2 = 34$, $\hat{p}_1 = 0.2843$, $\hat{p}_2 = 0.3333$,

$\hat{p} = 0.3088$, $\alpha = 0.05$, $Z_\alpha = 1.645$.

H_0: $p_2 \leq p_1$ against H_1: $p_2 > p_1$.

Test statistic: $Z = 0.7574$.

Conclusion: Since $Z = 0.7574 < 1.645$, do not reject H_0. That is, there is insufficient sample evidence to conclude that the vaccine is effective in preventing colds.

15. $n_1 = 11{,}000$, $n_2 = 11{,}000$, $x_1 = 104$, $x_2 = 189$, $\hat{p}_1 = 0.0095$,

$\hat{p}_2 = 0.0172$, $\hat{p} = 0.0133$, $\alpha = 0.05$, $Z_\alpha = 1.645$.

H_0: $p_2 \leq p_1$ against H_1: $p_2 > p_1$.

Test statistic: $Z = 4.9841$.

p-value $== p\{Z > 4.9841\} = 0.0$.

Conclusion: Since p-value = 0 (extremely small), reject H_0. That is, there is sufficient sample evidence to conclude that the proportion of heart attacks is less for the population that is taking a daily dose of aspirin.

REVIEW PROBLEMS

1. (a) $n = 1820$, $m = 1340$, $\overline{X} = 3480$, $\overline{Y} = 3260$, $S_x^2 = (9.2)^2$, $S_y^2 = (10.4)^2$,

$\alpha = 0.05$, $Z_{\alpha/2} = 1.96$.

H_0: $\mu_x = \mu_y$ against H_1: $\mu_x \neq \mu_y$

Test statistic: $Z = 616.7962$.

Conclusion: Since $|Z| = 616.7962 > 1.96$, reject H_0. That is, the data are consistent with the alternative hypothesis of the means not being equal at the 5% level of significance.

(b) p-value $= 2P\{Z > 616.7962\} = 0$.

3. $n_1 = 240$, $n_2 = 1200$, $x_1 = 13$, $x_2 = 78$, $\hat{p}_1 = 0.0542$, $\hat{p}_2 = 0.0650$,

$\hat{p} = 0.0632$, $\alpha = 0.05$, $Z_\alpha = 1.645$, $Z_{\alpha/2} = 1.96$.

(a) H_0: $p_2 = p_1$ against H_1: $p_2 \neq p_1$.

Test statistic: $Z = 0.6277$.

Conclusion: Since $Z = 0.6266 < 1.96$, do not reject H_0. That is, the data are not inconsistent with the null hypothesis that the proportion of knee injuries for players wearing conventional shoes to be equal to the proportion of those players wearing multicleated shoes at the 5% level of significance.

(b) p-value $= 2P\{Z \geq 0.6277\} = 2(1 - 0.7357) = 0.5286$.

(c) $H_0: p_2 \leq p_1$ against $H_1: p_2 > p_1$.

Test statistic: $Z = 0.6277$.

Conclusion: Since $Z = 0.6266 < 1.645$, do not reject H_0. That is, the data are not inconsistent with the null hypothesis that the proportion of knee injuries for players wearing conventional shoes to be the less than or equal to the proportion of those players wearing multicleated shoes at the 5% level of significance.

(d) p-value $= P\{Z \geq 0.6277\} = (1 - 0.7357) = 0.2643$. Thus, for

$\alpha \geq 0.2643$ the sample evidence will be strong or significant enough.

5. (a) $n = 15$, $m = 15$, $\overline{X} = 20.19$, $\overline{Y} = 17.58$, $S_x^2 = (3.62)^2$, $S_y^2 = (2.05)^2$,

$\alpha = 0.05$, $t_{28, \alpha/2} = 2.048$, $S_p^2 = (2.94)^2$.

$H_0: \mu_x = \mu_y$ against $H_1: \mu_x \neq \mu_y$

Test statistic: $T = (20.19 - 17.58)/(1.07) = 2.44$.

Conclusion: Since $|T| = 2.44 > 2.048$, reject H_0. That is, the data are consistent with the alternative hypothesis that the average heights for the cross-fertilized and self-fertilized Zea mays (corn) plants are different at the 5% level of significance.

Note: This approach was used since the treatments were applied to different experimental units.

(b) p-value $= 2P\{T \geq 2.44\} \approx 2(0.01) = 0.02$, df $= 28$.

7. $n_1 = 1184$, $n_2 = 399$, $x_1 = 93$, $x_2 = 20$, $\hat{p}_1 = 0.0785$, $\hat{p}_2 = 0.0501$,

$\hat{p} = 0.0714$, $\alpha = 0.05$, $Z_{\alpha/2} = 1.96$.

H_0: $p_2 = p_1$ against H_1: $p_2 \neq p_1$.

Test statistic: $Z = 1.9054$.

Conclusion: Since $Z = 1.9054 < 1.96$, do not reject H_0. That is, the data are not inconsistent with the null hypothesis that the proportion of both sexes dying from cancer is the same at the 5% level of significance.

9. $n_1 = 1728$, $n_2 = 657$, $x_1 = 685$, $x_2 = 257$, $\hat{p}_1 = 0.396$, $\hat{p}_2 = 0.39$,

$\hat{p} = 0.3950$, $\alpha = 0.05$, $Z_{\alpha/2} = 1.96$.

H_0: $p_1 = p_2$ against H_1: $p_1 \neq p_2$.

Test statistic: $Z = 0.2678$.

Conclusion: Since $Z| = 0.2678 < 1.96$, do not reject H_0. That is, the data are not inconsistent with the null hypothesis that the proportion of runs scored for the two given situations are the same at the 5% level of significance.

p-value $= 2P\{Z > 0.2678\} = 2(1 - 0.6064) = 0.7872$.

11. $n_1 = 100$, $n_2 = 99$, $x_1 = 53$, $x_2 = 57$, $\hat{p}_1 = 0.53$, $\hat{p}_2 = 0.5758$,

$\hat{p} = 0.5528$, $\alpha = 0.05$, $Z_{\alpha/2} = 1.96$.

H_0: $p_1 = p_2$ against H_1: $p_1 \neq p_2$.

Test statistic: $Z = -0.6497$.

Conclusion: Since $|Z = 0.6497 < 1.96$, do not reject H_0. That is, the data are not inconsistent with the null hypothesis that the proportion of games won by the home team are the same for both football and baseball at the 5% level of significance.

13. $n = 33$, $m = 33$, $\overline{X}_1 = 0.015$, $\overline{X}_2 = 0.006$, $S_1 = 0.004$, $S_2 = 0.006$,

$$\sqrt{\frac{S_1^2}{n} + \frac{S_2^2}{m}} = 0.0013.$$

H_0: $\mu_x = \mu_y$ against H_1: $\mu_x \neq \mu_y$

Test statistic: $Z = (0.015 - 0.006)/(0.0013) = 6.9231$.

p-value $= 2P\{Z > 6.9231\} \approx 0.0$.

Conclusion: Since p-value ≈ 0, reject H_0. That is, the mean blood lead levels for children of parents who work in a factory that uses lead and for those who do not, is not the same at the 5% level of significance.

Chapter 11 ANALYSIS OF VARIANCE

SECTION 11.2 - ONE-FACTOR ANALYSIS OF VARIANCE

PROBLEMS

1. (a) $\overline{X}_1 = 8$, $\overline{X}_2 = 14$, $\overline{X}_3 = 11$.

    (b) $\overline{\overline{X}} = (8 + 14 + 11)/3 = 11$.

    (c) $\overline{\overline{X}} = (5 + 9 + 12 + 6 + 13 + 12 + 20 + 11 + 8 + 12 + 16 + 8)/12 = 11$.

3. Below is a portion of a *Minitab* output for this problem.

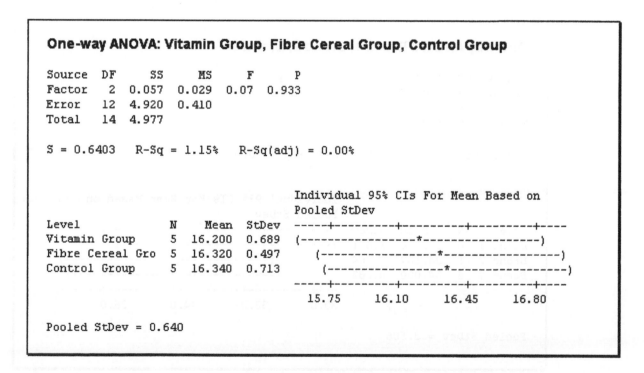

```
One-way ANOVA: Vitamin Group, Fibre Cereal Group, Control Group

Source   DF    SS      MS      F      P
Factor    2   0.057   0.029   0.07   0.933
Error    12   4.920   0.410
Total    14   4.977

S = 0.6403    R-Sq = 1.15%    R-Sq(adj) = 0.00%


                                  Individual 95% CIs For Mean Based on
                                  Pooled StDev
Level              N     Mean   StDev  -----+---------+---------+---------+----
Vitamin Group      5   16.200   0.689  (----------------*------------------)
Fibre Cereal Gro   5   16.320   0.497      (----------------*----------------)
Control Group      5   16.340   0.713       (----------------*----------------)
                                         -----+---------+---------+---------+----
                                         15.75     16.10     16.45     16.80

Pooled StDev = 0.640
```

From this output, $5\overline{S}^2 = 0.029$ and $\sum_{i=1}^{3} S_i^2 /3 = 0.410$. Numerator degrees of freedom = 2, denominator degrees of freedom = 12, $\alpha = 0.05$, $F_{2,\,12,\,0.05} = 3.89$.

H_0: $\mu_1 = \mu_2 = \mu_3$

H_1: at least one of the means is different from the rest.

Test Statistic: $F = 0.029/0.410 = 0.07$.

Conclusion: Since $F = 0.07 < 3.89$, do not reject H_0. That is, there is insufficient sample evidence to conclude that the means are different at the 5% level of significance. Thus, the data are consistent with the hypothesis that neither the vitamin nor fiber cereal affects the speed of the bicyclist.

Note: This result is confirmed by the p-value of 0.933 from the output Also, observe that the confidence intervals for the three populations overlap and hence you cannot claim that they are different.

5. Below is a portion of a *Minitab* output for this problem.

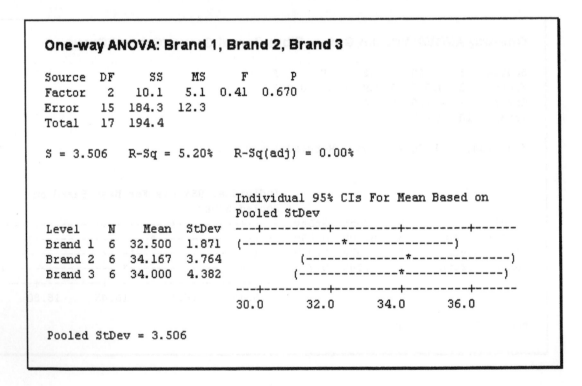

From this output, $6\overline{S}^2 = 5.1$ and $\sum_{i=1}^{3} S_i^2 /3 = 12.3$. Numerator degrees of freedom = 2, denominator degrees of freedom = 15, $\alpha = 0.05$, $F_{2,\,15,\,0.05} = 3.68$.

H_0: $\mu_1 = \mu_2 = \mu_3$

H_1: at least one of the means is different from the rest.

Test Statistic: F = 5.1/12.3 = 0.41.

Conclusion: Since F = 0.41 < 3.68, do not reject H_0. That is, there is insufficient sample evidence to conclude that the mean fat content for the three brands of processed meat are different at the 5% level of significance.

Note: This result is confirmed by the p-value of 0.670 from the output. Also, observe that the confidence intervals for the three populations overlap and hence one cannot claim that they are different.

7. Below is a portion of a *Minitab* output for this problem.

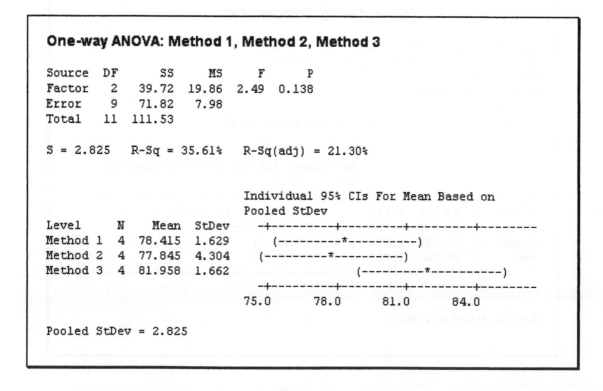

```
One-way ANOVA: Method 1, Method 2, Method 3

Source   DF      SS      MS     F      P
Factor    2    39.72   19.86  2.49  0.138
Error     9    71.82    7.98
Total    11   111.53

S = 2.825    R-Sq = 35.61%    R-Sq(adj) = 21.30%


                                Individual 95% CIs For Mean Based on
                                Pooled StDev
Level       N    Mean   StDev    -+---------+---------+---------+--------
Method 1    4  78.415   1.629         (---------*----------)
Method 2    4  77.845   4.304     (---------*----------)
Method 3    4  81.958   1.662                   (---------*----------)
                                  -+---------+---------+---------+--------
                                 75.0      78.0      81.0      84.0

Pooled StDev = 2.825
```

From this output, $4\overline{S}^2 = 19.86$ and $\sum_{i=1}^{3} S_i^2 /3 = 7.98$. Numerator degrees of freedom = 2, denominator degrees of freedom = 9, $\alpha = 0.05$,

$F_{2, 9, 0.05} = 4.26$.

H_0: $\mu_1 = \mu_2 = \mu_3$

H_1: at least one of the means is different from the rest.

Test Statistic: F = 19.86/7.98 = 2.49.

Conclusion: Since F = 2.49 < 4.26, do not reject H_0. That is, there is insufficient sample evidence to conclude that the mean readings for the three methods are different at the 5% level of significance.

Note: This result is confirmed by the p-value of 0.138 from the output.   Also, observe that the confidence intervals for the three populations overlap and hence one cannot claim that they are different.

9. Below is a portion of a *Minitab* output for this problem.

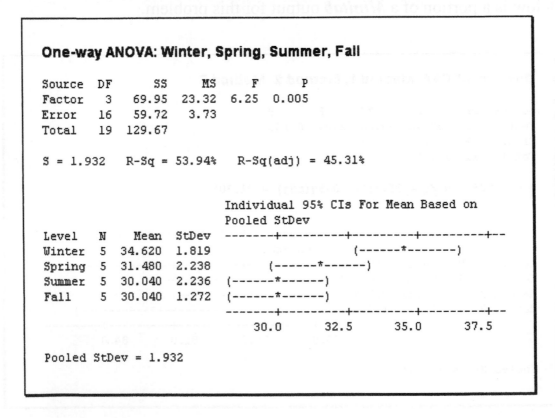

From this output, $5\overline{S}^2 = 23.32$ and $\displaystyle\sum_{i=1}^{4} S_i^2 /4 = 3.73$. Numerator degrees of freedom = 2, denominator degrees of freedom = 16, $\alpha = 0.05$,

$F_{3, 16, 0.05} = 3.24.$

$H_0: \mu_1 = \mu_2 = \mu_3 = \mu_4$

H_1: at least one of the means is different from the rest.

Test Statistic: $F = 23.32/3.73 = 6.25.$

Conclusion: Since $F = 6.25 > 3.24$, reject H_0. That is, there is sufficient sample evidence to conclude that the mean death rate (per 10,000 adults) are different for the various seasons at the 5% level of significance.

Note: This result is confirmed by the p-value of 0.005 from the output Also, observe that the confidence intervals for the four populations (seasons) do not all overlap and hence one can claim that the average death rates are different.

11. Below is a portion of a *Minitab* output for this problem.

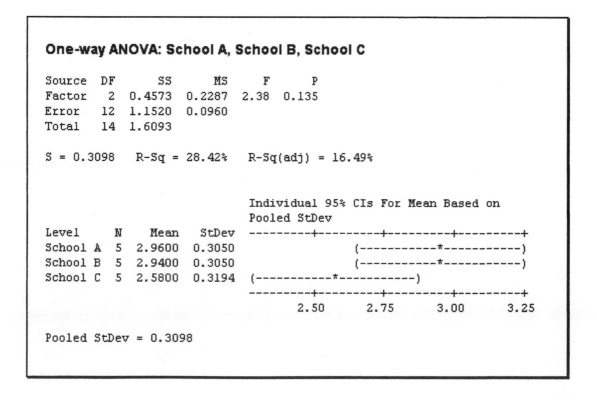

From this output, $5\overline{S}^2 = 0.2287$ and $\sum_{i=1}^{3} S_i^2 / 3 = 0.0960$. Numerator degrees of freedom = 2, denominator degrees of freedom = 12, $\alpha = 0.05$,

$F_{2, 12, 0.05} = 3.89$.

$H_0: \mu_1 = \mu_2 = \mu_3$

H_1: at least one of the means is different from the rest.

Test Statistic: $F = 0.2287/0.0960 = 2.38$.

Conclusion: Since $F = 2.38 < 3.89$, do not reject H_0. That is, there is insufficient sample evidence to conclude that the mean grade-point averages for the three schools are different at the 5% level of significance. Thus, there is not strong enough evidence to disprove the administrator's claim.

Note: This result is confirmed by the p-value of 0.135 from the output. Also, observe that the confidence intervals for the three populations overlap and hence you cannot claim that they are different.

SECTION 11.3 - TWO-FACTOR ANALYSIS OF VARIANCE: INTRODUCTION AND PARAMETER ESTIMATION

PROBLEMS

1. $X_{1.} = 83$, $X_{2.} = 75.33$, $X_{3.} = 65.33$, $X_{4.} = 65.33$, $X_{5.} = 55$, $X_{.1} = 69.6$,

   $X_{.2} = 66.4$, $X_{.3} = 70.4$, $X.. = 68.8$.

   $\hat{\mu} = 68.8$, $\hat{\alpha}_1 = 83 - 68.8 = 14.2$, $\hat{\alpha}_2 = 75.33 - 68.8 = 6.53$,

   $\hat{\alpha}_3 = 65.33 - 68.8 = -3.47$, $\hat{\alpha}_4 = 65.33 - 68.8 = -3.47$,

   $\hat{\alpha}_5 = 55 - 68.8 = -13.5$, $\hat{\beta}_1 = 69.6 - 68.8 = 0.8$,

   $\hat{\beta}_2 = 66.4 - 68.8 = -2.4$, $\hat{\beta}_3 = 70.4 - 68.8 = 1.6$.

3. $X_{1.} = 29.33$, $X_{2.} = 26.33$, $X_{3.} = 29.33$, $X_{.1} = 32$,

   $X_{.2} = 27.66$, $X_{.3} = 25.33$, $X.. = 28.33$.

   $\hat{\mu} = 28.33$, $\hat{\alpha}_1 = 29.33 - 28.33 = 1$, $\hat{\alpha}_2 = 26.33 - 28.33 = -2$,

   $\hat{\alpha}_3 = 29.33 - 28.33 = 1$, $\hat{\beta}_1 = 32 - 28.33 = 3.67$,

   $\hat{\beta}_2 = 27.66 - 28.33 = -0.67$, $\hat{\beta}_3 = 25.33 - 28.33 = -3$.

5. $X_{1.} = 13.39$, $X_{2.} = 10.08$, $X_{3.} = 11.08$, $X_{4.} = 10.15$, $X_{.1} = 10.35$,

   $X_{.2} = 10.49$, $X_{.3} = 10.61$, $X_{.4} = 13.25$, $X.. = 11.18$.

   (a) $\hat{\mu} = 11.18$.

   (b) $\hat{\alpha}_4 = 10.15 - 11.18 = -1.03$.

   (c) $\hat{\beta}_4 = 13.25 - 11.18 = 2.07$.

7. $X_{1.} = 4.2$, $X_{2.} = 5.0$, $X_{3.} = 7.3$, $X_{4.} = 5.13$, $X_{5.} = 5$,

$X_{.1} = 4.14$, $X_{.2} = 5.14$, $X_{.3} = 6.5$, $X_{..} = 5.36$.

$\hat{\mu} = 5.36$, $\hat{\alpha}_1 = 4.2 - 5.36 = -1.16$, $\hat{\alpha}_2 = 5 - 5.36 = -0.36$,

$\hat{\alpha}_3 = 7.3 - 5.36 = 1.94$, $\hat{\alpha}_4 = 5.13 - 5.36 = -0.23$,

$\hat{\alpha}_5 = 5 - 5.36 = -0.36$, $\hat{\beta}_1 = 4.14 - 5.36 = -1.22$,

$\hat{\beta}_2 = 5.14 - 5.36 = -0.22$, $\hat{\beta}_3 = 6.5 - 5.36 = 114$.

9. From problem number 8, the rectangular array of rows and columns is given below:

$$\begin{bmatrix} 5 & 9 & 13 & 17 \\ 6 & 10 & 14 & 18 \\ 7 & 11 & 15 & 19 \end{bmatrix}$$

(a) $\sum_{j=1}^{4} x_{1j} = (5 + 9 + 13 + 17) = 44$.

(b) $\sum_{j=1}^{4} x_{2j} = (6 + 10 + 14 + 18) = 48$.

(c) $\sum_{j=1}^{4} x_{3j} = (7 + 11 + 15 + 19) = 52$.

(d) $\sum_{j=1}^{4} x_{4j} = (44 + 48 + 52) = 144$.

SECTION 11.4 - TWO-FACTOR ANALYSIS OF VARIANCE: TESTING HYPOTHESES

PROBLEMS

1. Below is a partial *Minitab* output for this problem.

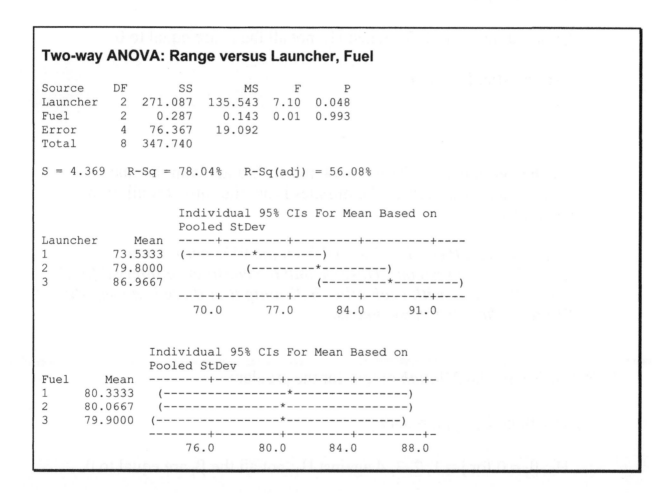

```
Two-way ANOVA: Range versus Launcher, Fuel

Source    DF      SS        MS      F      P
Launcher   2   271.087   135.543   7.10   0.048
Fuel       2     0.287     0.143   0.01   0.993
Error      4    76.367    19.092
Total      8   347.740

S = 4.369   R-Sq = 78.04%   R-Sq(adj) = 56.08%


                      Individual 95% CIs For Mean Based on
                      Pooled StDev
Launcher     Mean    -----+---------+---------+---------+----
1          73.5333   (---------*---------)
2          79.8000             (---------*---------)
3          86.9667                       (---------*---------)
                     -----+---------+---------+---------+----
                        70.0      77.0      84.0      91.0


                      Individual 95% CIs For Mean Based on
                      Pooled StDev
Fuel      Mean     --------+---------+---------+---------+-
1       80.3333    (----------------*----------------)
2       80.0667    (---------------*----------------)
3       79.9000    (----------------*---------------)
                   --------+---------+---------+---------+-
                        76.0      80.0      84.0      88.0
```

(a) $\alpha = 0.05$, $F_{2, 4, 0.05} = 6.94$.

H_0: $\alpha_i = 0$ for i = 1, 2, 3 against H_1: not all the α_i are equal to 0.

Test Statistic: F = 0.01.

Conclusion: Since $F = 0.01 < 6.94$, do not reject H_0 and claim that the mean distance traveled by the missiles is the same for all of the types of fuels.

Note: This result can be observed from the 95% confidence intervals for the systems. Observe that the confidence interval for Launcher 3 does not overlap with the interval for Launcher 1.

(b) $\alpha = 0.05$, $F_{2, 4, 0.05} = 6.94$.

H_0: $\beta_j = 0$ for $i = 1, 2, 3$ against H_1: not all the β_j are equal to 0.

Test Statistic: $F = 7.10$.

Conclusion: Since $F = 7.09 > 6.94$, reject H_0 and claim that the mean distance traveled by the missiles is not the same for all of the launchers.

Note: This result can be observed from the 95% confidence intervals for the systems. Observe that the confidence interval for the Fuels all overlap with each other indicating that there is no significant difference between these means.

3. Below is a partial *Minitab* output for the problem.

(a) $\alpha = 0.05$, $F_{3, 12, 0.05} = 3.49$.

H_0: $\beta_j = 0$ for $j = 1, 2, 3, 4$ against H_1: not all the β_j are equal to 0.

Test Statistic: $F = 3.72$.

Conclusion: Since $F = 3.72 > 3.49$, reject H_0. That is, there is sufficient sample evidence to claim that the mean yield for the different fields are not the same at the 5% level of significance.

Observe that the confidence interval for field number 4 does not overlap with the confidence interval for field number 3.

Two-way ANOVA: Yield versus Variety, Field

```
Source    DF        SS       MS      F      P
Variety    4    10234.3  2558.58   4.47   0.019
Field      3     6379.7  2126.58   3.72   0.042
Error     12     6868.5   572.38
Total     19    23482.6

S = 23.92    R-Sq = 70.75%    R-Sq(adj) = 53.69%


                       Individual 95% CIs For Mean Based on
                       Pooled StDev
Variety    Mean    -+---------+---------+---------+---------
1         335.25            (---------*-------)
2         386.75                        (--------*--------)
3         338.25             (---------*-------)
4         322.00      (-------*--------)
5         332.00          (---------*-------)
                     -+---------+---------+---------+---------
                     300       330       360       390


                       Individual 95% CIs For Mean Based on
                       Pooled StDev
Field    Mean    ----+---------+---------+---------+-----
1        345.4                (--------*--------)
2        350.0                (--------*--------)
3        362.2                     (--------*--------)
4        313.8    (---------*--------)
                 ----+---------+---------+---------+-----
                 300       325       350       375
```

(b) $\alpha = 0.05$, $F_{4, 12, 0.05} = 3.26$.

H_0: $\alpha_i = 0$ for $i = 1, 2, 3, 4, 5$ against H_1: not all the α_i are equal to 0.

Test Statistic: $F = 3.72$.

Test Statistic: $F = 4.47$.

Conclusion: Since $F = 4.47 > 3.26$, reject H_0. That is, there is sufficient sample evidence to claim that the mean yield for the different oat variety are not the same at the 5% level of significance.

*Note: These results can be observed from the 95% confidence
intervals/or the variables field and variety.*

5. Below is a partial *Minitab* output for this problem

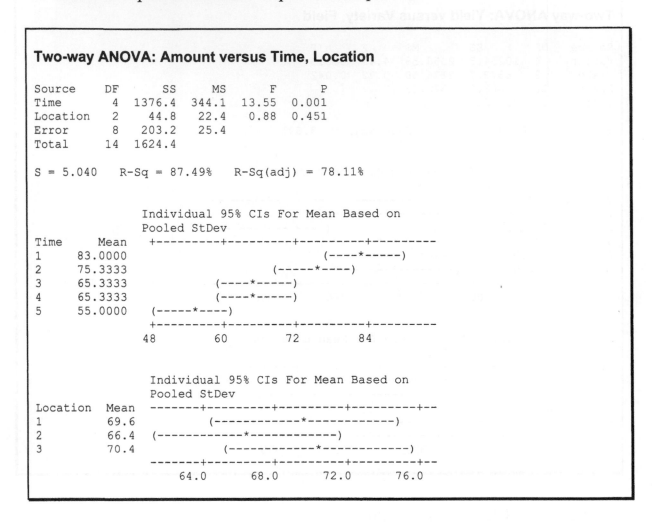

```
Two-way ANOVA: Amount versus Time, Location

Source      DF      SS      MS      F       P
Time        4    1376.4   344.1   13.55   0.001
Location    2      44.8    22.4    0.88    0.451
Error       8     203.2    25.4
Total      14    1624.4

S = 5.040    R-Sq = 87.49%    R-Sq(adj) = 78.11%


                        Individual 95% CIs For Mean Based on
                        Pooled StDev
Time     Mean    +---------+---------+---------+---------
1       83.0000                             (----*-----)
2       75.3333                     (-----*----)
3       65.3333          (----*-----)
4       65.3333          (----*-----)
5       55.0000    (-----*----)
                  +---------+---------+---------+---------
                  48        60        72        84


                        Individual 95% CIs For Mean Based on
                        Pooled StDev
Location  Mean   -------+---------+---------+---------+--
1        69.6              (------------*------------)
2        66.4     (------------*------------)
3        70.4          (------------*------------)
                 -------+---------+---------+---------+--
                        64.0      68.0      72.0      76.0
```

(a) $\alpha = 0.05$, $F_{4, 8, 0.05} = 3.84$.

H_0: $\alpha_i = 0$ for i = 1, 2, 3, 4, 5 against H_1: not all the α_i are equal to 0.

Test Statistic: F = 13.55.

Conclusion: Since F = 13.55 > 3.84, reject H_0. That is, there
is sufficient sample evidence to claim that the mean air pollution
(paniculate matter) for the different times are not the same at the 5%
level of significance.

(b) $\alpha = 0.05$, $F_{2, 8, 0.05} = 4.46$.

H_0: $\beta_j = 0$ for j = 1, 2, 3 against H_1: not all the β_j are equal to 0.

Test Statistic: F = 0.88.

Conclusion: Since F = 0.88 < 4.46, do not reject H_0. That is, there is insufficient sample evidence to claim that the mean air pollution (particulate matter) for the different times are not the same at the 5% level of significance.

Note: These results can be observed from the 95% confidence intervals for the variables location and time.

7. Below is a partial *Minitab* output for this problem.

(a) $\alpha = 0.05$, $F_{4, 20, 0.05} = 2.87$.

H_0: $\alpha_i = 0$ for i = 1, 2,..., 5 against Hi: not all the α_i are equal to 0.

Test Statistic: F = 21.45.

Conclusion: Since F = 21.45 > 2.87, reject H_0. That is, there is sufficient sample evidence to claim that the mean percentage of smokers for the different years are not the same at the 5% level of significance.

(b) $\alpha = 0.05$, $F_{5, 20, 0.05} = 2.71$.

H_0: $\beta_j = 0$ for j = 1, 2,.... 6 against H_1: not all the β_j are equal to 0.

Test Statistic: F = 29.8.

Conclusion: Since F = 29.8 > 2.71, reject H_1. That is, there is sufficient sample evidence to claim that the mean percentage of smokers for the different age groups are not the same at the 5% level of significance.

Note: These results can be observed from the 95% confidence intervals/or the variables year and age group.

```
Two-way ANOVA: Smokers versus Year, Age

Source   DF       SS        MS       F       P
Year      4    645.87   161.467   21.45   0.000
Age       5   1121.47   224.293   29.80   0.000
Error    20    150.53     7.527
Total    29   1917.87

S = 2.743   R-Sq = 92.15%   R-Sq(adj) = 88.62%


                    Individual 95% CIs For Mean Based on
                    Pooled StDev
Year      Mean   ---------+---------+---------+---------+
1       40.5000                            (----*----)
2       32.1667           (---*----)
3       29.8333        (----*---)
4       28.5000     (----*----)
5       27.6667   (---*----)
                 ---------+---------+---------+---------+
                       30.0      35.0      40.0      45.0


                   Individual 95% CIs For Mean Based on Pooled
                   StDev
Age  Mean       -+---------+---------+---------+--------
1    29.4                     (---*---)
2    38.8                            (----*---)
3    37.0                          (----*---)
4    33.6                      (---*---)
5    31.6                    (----*---)
6    20.0        (---*----)
                -+---------+---------+---------+--------
              18.0      24.0      30.0      36.0
```

9. Below is a partial *Minitab* output for this problem.

Two-way ANOVA: Rates versus Year, Industry

```
Source      DF       SS       MS       F       P
Year         2  18.0520  9.02600   28.87   0.000
Industry     4  18.7827  4.69567   15.02   0.001
Error        8   2.5013  0.31267
Total       14  39.3360

S = 0.5592    R-Sq = 93.64%    R-Sq(adj) = 88.87%


                  Individual 95% CIs For Mean Based on
                  Pooled StDev
Year  Mean   ----+---------+---------+---------+-----
1     4.14   (----*-----)
2     5.14             (----*-----)
3     6.80                                 (-----*-----)
             ----+---------+---------+---------+-----
                4.0       5.0       6.0       7.0


                  Individual 95% CIs For Mean Based on
                  Pooled StDev
Industry    Mean    -+---------+---------+---------+--------
1        4.20000    (------*-----)
2        4.96667            (-----*------)
3        7.50000                              (------*-----)
4        5.13333             (-----*-----)
5        5.00000            (------*-----)
                    -+---------+---------+---------+--------
                   3.6       4.8       6.0       7.2
```

(a) $\alpha = 0.05$, $F_{4, 8, 0.05} = 3.84$.

 H_0: $\alpha_i = 0$ for $i = 1, 2, ..., 5$ against Hi: not all the α_i are equal to 0.

 Test Statistic: $F = 15.02$.

Conclusion: Since F = 15.02 > 3.84, reject H_0. That is, there
is sufficient sample evidence to claim that the mean employment rate
for the different industries are not the same at the 5% level of
significance.

(b) $\alpha = 0.05$, $F_{2, 8, 0.05} = 4.46$.

H_0: $\beta_j = 0$ for i = 1, 2, 3 against H_1: not all the β_j are equal to 0.

Test Statistic: F = 28.87.

Conclusion: Since F = 28.87 > 4.46, reject H_0. That is, there
is sufficient sample evidence to claim that the mean employment rate
for the different years are not the same at the 5% level of significance.

*Note: These results can be observed from the 95% confidence
intervals/or the variables industry and year.*

REVIEW PROBLEMS

1. $10\overline{S}^2 = 19390$ and $\sum_{i=1}^{3} S_i^2 / 3 = 471.3333$. Numerator degrees of freedom = 2, denominator degrees of freedom = 27, $\alpha = 0.05$,

   $F_{2, 27, 0.05} \approx 3.39$ (3.32)

   H_0: $\mu_1 = \mu_2 = \mu_3$

   H_1: at least one of the means is different from the rest.

   Test Statistic: $F = 19390/471.333 = 41.1386$.

   Conclusion: Since $F = 41.1386 > 3.39$, reject H_0. That is, there is sufficient sample evidence to conclude that the average daily production for the three plants are different at the 5% level of significance.

3. Below is a portion of a *Minitab* output for this problem.

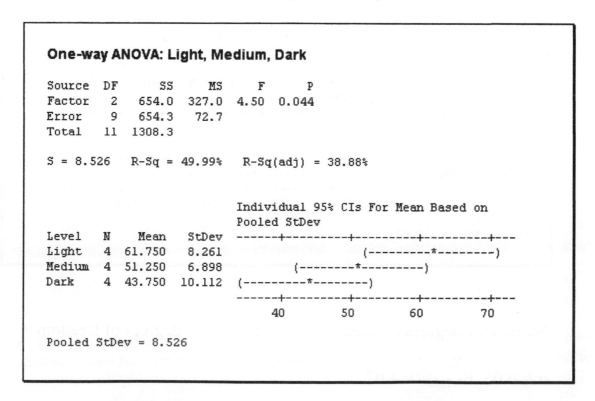

```
One-way ANOVA: Light, Medium, Dark

Source  DF     SS     MS     F      P
Factor   2   654.0  327.0  4.50  0.044
Error    9   654.3   72.7
Total   11  1308.3

S = 8.526   R-Sq = 49.99%   R-Sq(adj) = 38.88%


                            Individual 95% CIs For Mean Based on
                            Pooled StDev
Level    N    Mean   StDev  ------+---------+---------+---------+---
Light    4  61.750   8.261                        (---------*--------)
Medium   4  51.250   6.898              (--------*---------)
Dark     4  43.750  10.112  (---------*--------)
                            ------+---------+---------+---------+---
                              40        50        60        70


Pooled StDev = 8.526
```

Numerator degrees of freedom = 2, denominator degrees of freedom = 9,

$\alpha = 0.05$, $F_{2, 9, 0.05} = 4.26$.

H_0: $\mu_1 = \mu_2 = \mu_3$

H_1: at least one of the means is different from the rest.

Test Statistic: F = 4.5.

Conclusion: Since F = 4.5 > 4.26, reject H_0. That is, there is sufficient sample evidence to conclude that the average sensitivity to pain for the three hair colors are different at the 5% level of significance.

5. Below is a portion of a *Minitab* output for this problem.

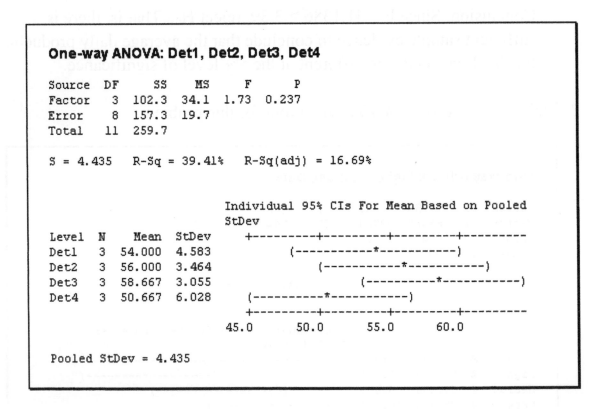

```
One-way ANOVA: Det1, Det2, Det3, Det4

Source   DF      SS      MS      F      P
Factor    3    102.3    34.1   1.73   0.237
Error     8    157.3    19.7
Total    11    259.7

S = 4.435     R-Sq = 39.41%     R-Sq(adj) = 16.69%


                                Individual 95% CIs For Mean Based on Pooled
                                StDev
Level  N     Mean    StDev      +---------+---------+---------+---------
Det1   3   54.000   4.583              (------------*------------)
Det2   3   56.000   3.464                 (------------*------------)
Det3   3   58.667   3.055                    (----------*-----------)
Det4   3   50.667   6.028        (----------*------------)
                                +---------+---------+---------+---------
                               45.0      50.0      55.0      60.0

Pooled StDev = 4.435
```

Numerator degrees of freedom = 3, denominator degrees of freedom = 8,

$\alpha = 0.05$, $F_{3, 8, 0.05} = 4.07$.

$H_0: \mu_1 = \mu_2 = \mu_3 = \mu_4$

H_1: at least one of the means is different from the rest.

Test Statistic: $F = 1.73$.

Conclusion: Since $F = 1.73 < 4.07$, do not reject H_0. That is, there is insufficient sample evidence to conclude that the detergents are not equally effective at the 5% level of significance.

7. Below is a portion of a *Minitab* output for this problem.

```
Two-way ANOVA: Days versus Woman, Dye

Source   DF      SS       MS       F       P
Woman    2     921.5   460.750   24.57   0.001
Dye      3      43.0    14.333    0.76   0.554
Error    6     112.5    18.750
Total   11    1077.0

S = 4.330    R-Sq = 89.55%    R-Sq(adj) = 80.85%


                     Individual 95% CIs For Mean Based on Pooled
                     StDev
Woman   Mean     -+---------+---------+---------+--------
1       20.75     (-------*------)
2       28.75               (------*------)
3       42.00                              (------*-----)
                 -+---------+---------+---------+--------
                 16.0      24.0      32.0      40.0


                     Individual 95% CIs For Mean Based on
                     Pooled StDev
Dye     Mean     --------+---------+---------+---------+-
1       27.3333   (-----------*-----------)
2       32.3333             (-----------*-----------)
3       31.0000         (-----------*-----------)
4       31.3333          (-----------*-----------)
                 --------+---------+---------+---------+-
                        25.0      30.0      35.0      40.0
```

(a) $\alpha = 0.05$, $F_{3, 6, 0.05} = 4.76$.

H_0: $\beta_j = 0$ for $i = 1, 2, 3, 4$ against H_1: not all the β_j are equal to 0.

Test Statistic: $F = 0.76$.

Conclusion: Since $F = 0.76 < 4.76$, do not reject H_0. That is, there is insufficient sample evidence to claim that the mean lasting power for the four dyes are not the same at the 5% level of significance.

(b) $X_{1.} = 20.75$, $X_{2.} = 28.75$, $X_{3.} = 42$, $X_{.1} = 27.3$,

$X_{.2} = 32.3$, $X_{.3} = 31$, $X_{.4} = 31.3$, $X_{..} = 30.5$.

$\hat{\mu} = 30.5$, $\hat{\alpha}_1 = 20.75 - 30.5 = -9.75$, $\hat{\alpha}_2 = 28.75 - 30.5 = -1.75$,

$\hat{\alpha}_3 = 42 - 30.5 = 11.51$, $\hat{\beta}_1 = 27.3 - 30.5 = -3.2$,

$\hat{\beta}_2 = 32.3 - 30.5 = 1.8$, $\hat{\beta}_3 = 31 - 30.5 = 0.5$, $\hat{\beta}_4 = 31.3 - 30.5 = 0.8$

Thus, $E[X_{22}] = 30.5 + (-1.75) + 1.8 = 30.55$.

(c) $\alpha = 0.05$, $F_{3, 6, 0.05} = 4.76$.

H_0: $\alpha_i = 0$ for $i = 1, 2, 3$ against H_1: not all the α_i are equal to 0.

Test Statistic: $F = 24.57$.

Conclusion: Since $F = 24.57 > 4.76$, reject H_0. That is, there is sufficient sample evidence to claim that the mean lasting power does depend on which woman is being tested at the 5% level of significance.

Note: These results can be observed from the 95% confidence intervals for the variables dye and woman.

9. Below is a portion of a *Minitab* output for this problem.

```
Two-way ANOVA: Deaths versus Year, Season

Source   DF        SS        MS      F      P
Year      4      9.507    2.3768   0.57   0.691
Season    3     69.949   23.3165   5.57   0.012
Error    12     50.213    4.1844
Total    19    129.669

S = 2.046    R-Sq = 61.28%    R-Sq(adj) = 38.69%


                      Individual 95% CIs For Mean Based on Pooled
                      StDev
Year     Mean        +---------+---------+---------+---------
1        31.725                 (-----------*----------)
2        30.250       (-----------*----------)
3        31.675                 (----------*----------)
4        31.750                 (----------*----------)
5        32.325                    (-----------*----------)
                      +---------+---------+---------+---------
                    28.0      30.0      32.0      34.0


                      Individual 95% CIs For Mean Based on
                      Pooled StDev
Season   Mean        --------+---------+---------+---------+-
1        34.62                            (------*-------)
2        31.48                 (-------*-------)
3        30.04        (-------*-------)
4        30.04        (-------*-------)
                      --------+---------+---------+---------+-
                            30.0      32.5      35.0      37.5
```

(a) $\alpha = 0.05$, $F_{4, 12, 0.05} = 3.26$.

H_0: $\alpha_i = 0$ for $i = 1, 2, 3, 4, 5$ against H_1: not all the α_i are equal to 0.

Test Statistic: $F = 0.57$.

Conclusion: Since $F = 0.57 < 3.26$, do not reject H_0. That is, there is insufficient sample evidence to claim that the mean number of deaths for the different years are not the same at the 5% level of significance.

(b) $\alpha = 0.05$, $F_{3, 12, 0.05} = 3.49$.

H_0: $\beta_j = 0$ for $i = 1, 2, 3, 4$ against H_1: not all the β_j are equal to 0.

Test Statistic: $F = 5.57$.

Conclusion: Since $F = 5.57 > 3.49$, reject H_0. That is, there is sufficient sample evidence to claim that the mean number of deaths for the different seasons are not the same at the 5% level of significance.

Note: These results can be observed from the 95% confidence intervals for the variables year and season.

Chapter 12   LINEAR REGRESSION

SECTION 12.2 - SIMPLE LINEAR REGRESSION MODEL

PROBLEMS

1. (a) Let Y = percentage yield of a laboratory experiment.
        Let x = temperature at which the experiment was conducted.

       The scatter diagram is shown below.

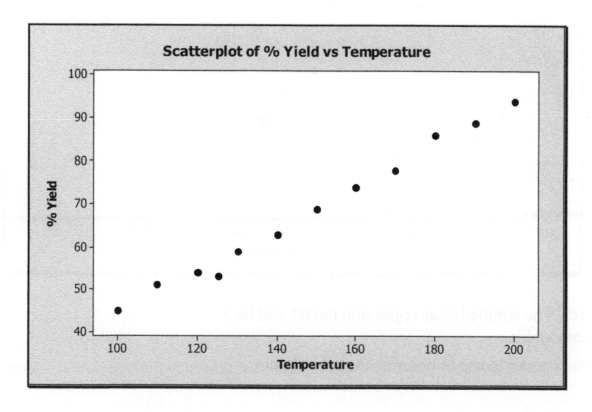

    (b) Based on the scatter plot, one can observe a straight line pattern.
         Thus, a simple linear regression model will be appropriate

to describe
   the relationship.

3. (a) Let Y = speed (response variable).

      Let x = traffic density (input variable).

   (b) The scatter diagram for the speed versus traffic density is shown
      below.

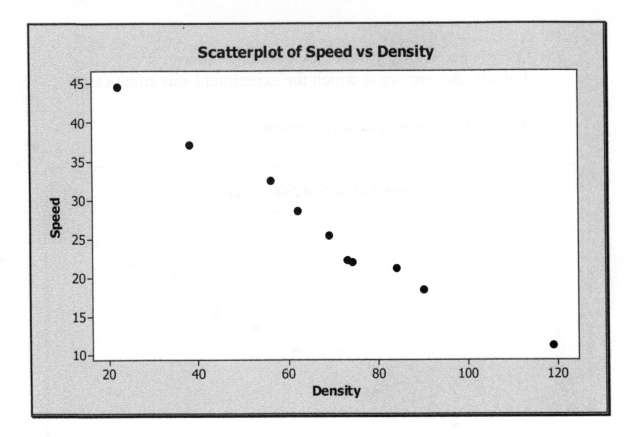

   (c) The simple linear regression model will be a reasonable
      model to use to describe the relationship.

5. (a) Let Y = miles of use (mileage).

      Let x = tire pressure.

The scatter diagram miles of use versus tire pressure is
shown
below.

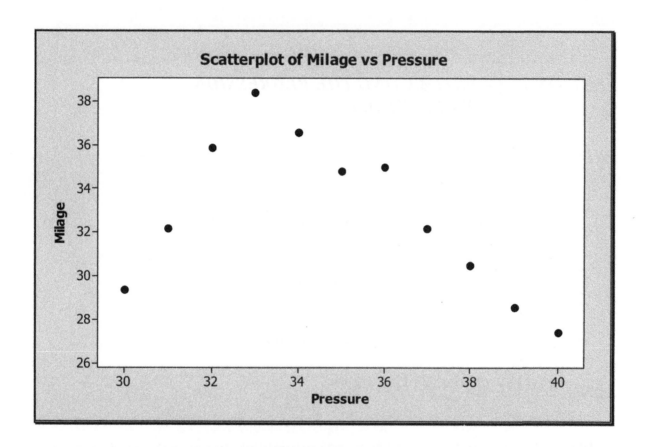

(b) The simple linear regression model is not a
reasonable model
     to use to describe the relationship between
mileage and tire
     pressure. The relationship is nonlinear.

SECTION 12.3 - ESTIMATING THE REGRESSION PARAMETERS

PROBLEMS

1. $n = 4$, $\displaystyle\sum_{i=1}^{4} x = 11$, $\displaystyle\sum_{i=1}^{4} Y = 31$, $\displaystyle\sum_{i=1}^{4} (xY) = 102$, $\displaystyle\sum_{i=1}^{4} x^2 = 39$,

   $\displaystyle\sum_{i=1}^{4} Y^2 = 273$, $S_{xY} = 16.75$, $S_{xx} = 8.75$, $S_{YY} = 32.75$.

   So, $\hat{\beta} = S_{xY}/ S_{xx} = 16.75/8.75 = 1.9143$, and

   $\hat{\alpha} = (31/4) - (1.9143)(11/4) = 2.4857$.

   (a)

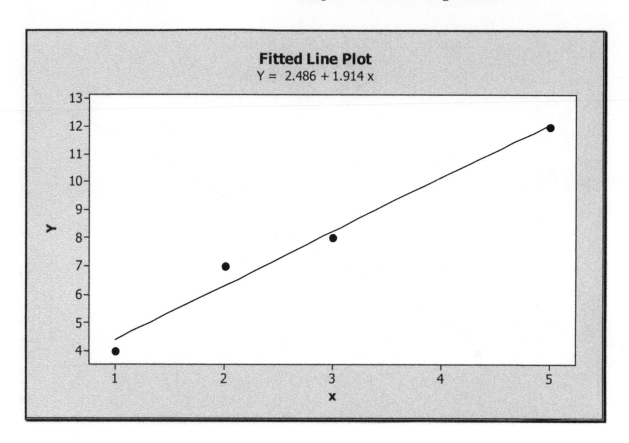

Fitted Line Plot
Y = 2.486 + 1.914 x

(b) $\hat{\alpha} = 4.971$, $\hat{\beta} = 1.914$. Note: When the data values are doubled,

the value for the Y-intercept is doubled but the slope remains the

same.

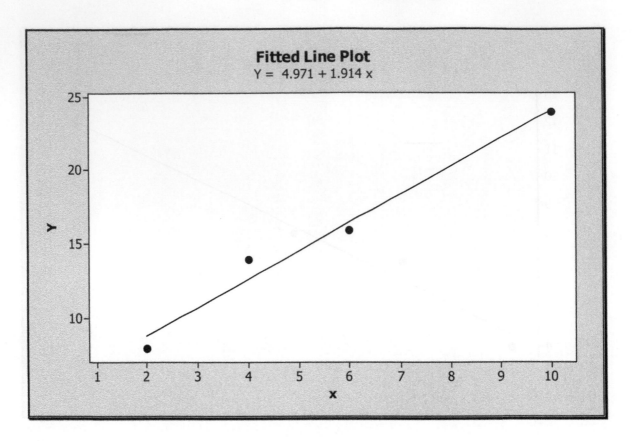

3. (a) The scatter diagram for damage versus distance is shown below.

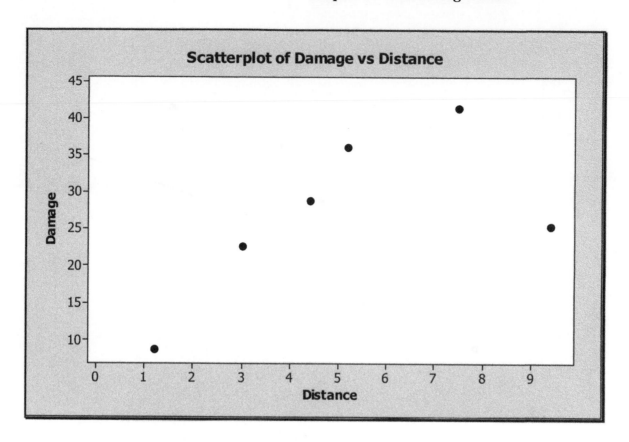

(b) The diagram below shows an approximate relationship between

the distance and damage. The end-coordinates used were (1, 15)

and (9.5, 35). The estimated line from these coordinates is

Y = 2.35x + 12.65.

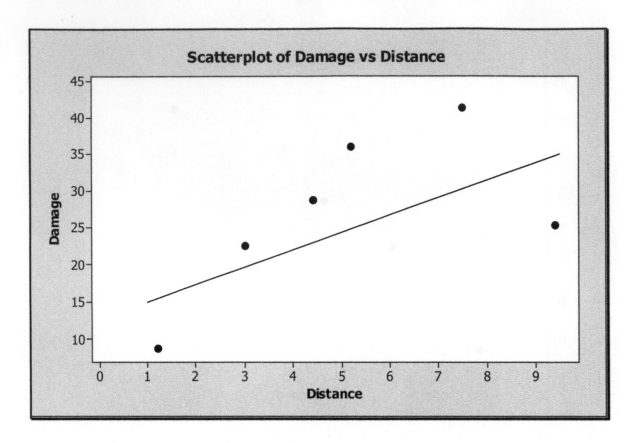

(c) The following diagram shows the estimated regression line.

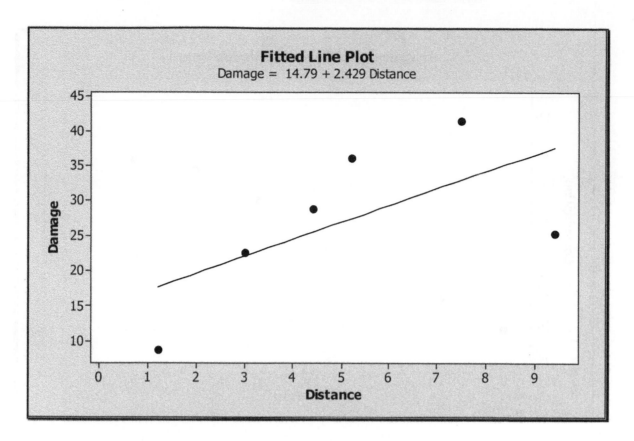

Fitted Line Plot
Damage = 14.79 + 2.429 Distance

The simple linear regression model is not a reasonable model to use to describe the relationship between damage and distance. A better model would be a nonlinear model.

5. (a) Let Y = amount of poultry products consumption (in pounds).

Let x = year.

The scatter diagram is shown below.

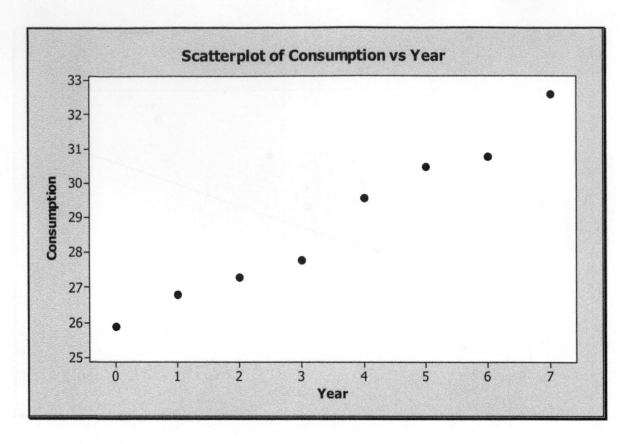

(b) The estimated regression line is $Y = 25.65 + 0.9321$ year.

(c) A plot of the estimated regression line on the scatter diagram is given below.

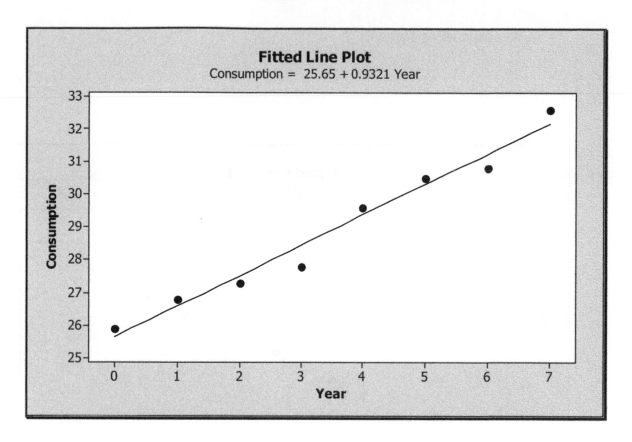

Fitted Line Plot

Consumption = 25.65 + 0.9321 Year

(d) The predicted amount of poultry product consumed in 1994 is

$$Y = 25.65 + 0.9321(-1) = 24.7179 \text{ (pounds)}.$$

(e) The predicted of poultry product consumed in 2004 is

$$Y = 25.65 + 0.9321(9) = 34.0389 \text{ (pounds)}.$$

7. (a) Let Y = world production of newsprint (in millions of metric tons).

Let x = world production of wood pulp (in millions of metric tons).

The scatter diagram and the estimated regression line are displayed in the following figure.

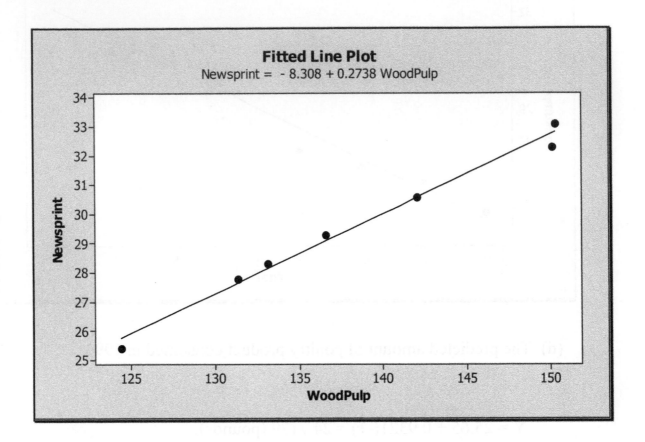

Fitted Line Plot
Newsprint = - 8.308 + 0.2738 WoodPulp

(b) The predicted amount of newsprint when 146.0 million

metric tons of wood pulp is produced is

Y = -8.308 + 0.2738(146) = 31.6668 metric tons.

(c) Let x = world production of newsprint (in millions of metric tons).

Let Y = world production of wood pulp (in millions of metric

tons).

The scatter diagram and the estimated regression line are displayed

in the following figure.

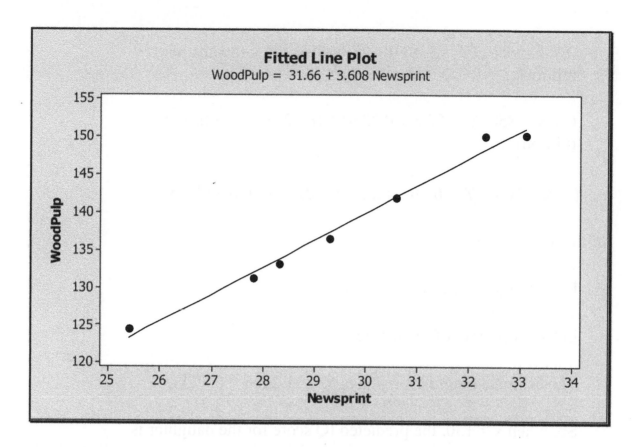

(d) The predicted amount of wood pulp when 32 million

metric tons of newsprint is produced is

Y = 31.66 + 3.608(32) = 147.116 metric tons.

9. The assignment of the eight workers to the training times should be made at random.

11. (a) Let Y = men's death rate from lung cancer in 1950 (per

million).

Let x = 1930 per capita consumption of cigarettes.

The estimated regression line is Y = 67.6 + 0.22 8x.

(b) x = 600, Y = 67.6 + 0.228(600) = 204.4 (deaths per million).

(c) x = 850, Y = 67.6 + 0.228(850) = 261.4 (deaths per million).

(d) x = 1000, Y = 67.6 + 0.228(1000) - 295.6 (deaths per

million).

13. Let Y = IQ score of the daughter.

Let x = IQ score of the mother.

The estimated regression line is Y = 21.2266 + 0.774x.

So, when x = 130, the predicted IQ score for the daughter is

Y = 21.2266 + 0.774(130) = 121.8466.

15. Let Y = proportion of coal miners who exhibit symptoms of pneumoconiosis.

Let x = number of years of working in the coal mines.

From the diagram below, the estimated regression line is

Y = -0.09913 + 0.009521x.

So, when x = 42, the predicted proportion of coal miners who exhibit symptoms of pneumoconiosis is

Y = -0.09913 + 0.009521(42) = 0.300752.

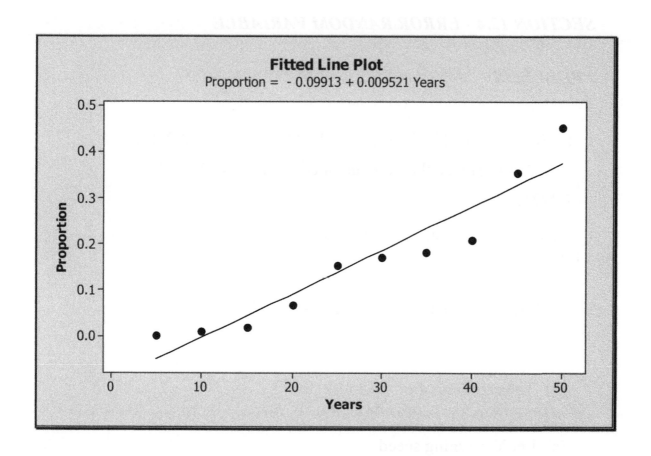

SECTION 12.4 - ERROR RANDOM VARIABLE

PROBLEMS

1. $n = 12$, $S_{xx} = 11{,}572.9$, $S_{YY} = 2{,}922.92$, $S_{xY} = 5{,}792.92$, SS_R $= 23.2214$. Hence the estimate of $\sigma^2 = 23.2214/(12 - 2) = 2.3221$.

3. (a) $n = 3$, $S_{xx} = 200$, $S_{YY} = 134$, $S_{xY} = 160$,

$$SS_R = [(200)(134) - 160^2]/200 = 6.$$

(b) The estimate of $\sigma^2 = 6/(3 - 2) = 6$.

(c) Let Y = typing speed.

Let x = temperature setting.

The estimated regression line is $Y = 24 + 0.8x$.

Thus, when $x = 65$, then the predicted typing speed is

$$Y = 24 + 0.8(65) = 76.$$

5. $n = 7$, $S_{xx} = 25.2143$, $S_{YY} = 0.0275$, $S_{xY} = 0.8214$, $SS_R = 0.00074$. Hence

the estimate of $\sigma^2 = 0.00074/(7 - 2) = 0.00015$.

7. $n = 11$, $S_{xx} = 1,432,255$, $S_{YY} = 137,473$, $S_{xY} = 327,182$,

$SS_R = 62,732.0767$. Hence the estimate of $\sigma^2 =$
$62,732.0767/(11 - 2)$

$= 6,970.2307$.

SECTION 12.5 - TESTING THE HYPOTHESIS THAT $\beta = 0$

PROBLEMS

1. $n = 4$, $\gamma = 0.05$, $t_{2,\,0.025} = 4.403$, $S_{xx} = 53$, $S_{YY} = 2$,

$S_{xY} = -2$, $SS_R = 1.9245$, $\sqrt{(n-2)S_{xx}/SS_R} = 7.4215$,

$\overset{\wedge}{\alpha} = 7.221$, $\overset{\wedge}{\beta} = -0.0377$.

$H_0: \beta = 0$ against $H_1: \beta \neq 0$.

TS: $T = [\sqrt{(n-2)S_{xx}/SS_R}] \times \overset{\wedge}{\beta} = (7.4215)(-0.0377) = -0.2798$.

Conclusion: Since $|T| = 0.2798 < 4.403$, do not reject H_0.
That is,
there is insufficient sample evidence to claim that the slope of the
regression line is not equal to zero at the 5% level of

significance.

3. $n = 11$, $\gamma = 0.05$, $t_{9,\,0.025} = 2.262$, $S_{xx} = $
26,318,

   $S_{YY} = 105.062$, $S_{xY} = 1104.5$, $SS_R = $
58.709,

$$\sqrt{(n-2)S_{xx}/SS_R} = 63.5178,\ \hat{\alpha} = 47.744,\ \hat{\beta} = $$
0.04197.

   H_0: $\beta = 0$ against H_1: $\beta \neq 0$.

   TS: $T = [\sqrt{(n-2)S_{xx}/SS_R}\ ]\times \hat{\beta} = (63.5178)(0.04197) = $
      2.6658.

Conclusion: Since $|T| = 2.6658 > 2.262$, reject H_0. That is,
there is sufficient sample evidence to claim that the number of
sunspots
and the number of automobile deaths are related at the 5% level of
significance.

5. (a) The scatter diagram of number of cigarettes smoked
versus
      death rate from bladder cancer is shown below.

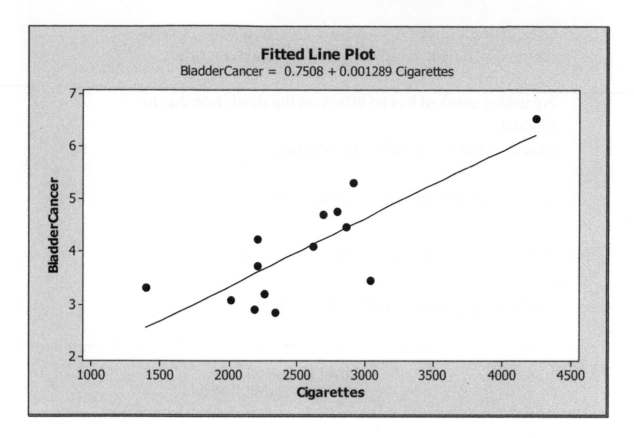

Fitted Line Plot
BladderCancer = 0.7508 + 0.001289 Cigarettes

(b) The estimated regression line is Y = 0.7508 + 0.001289x.

(c) n = 14, γ = 0.05, $t_{12,\,0.025}$ = 2.179, S_{xx} = 5,510,467,

S_{YY} = 14.2305, S_{xY} = 7103.32, SS_R = 5.0739,

$\sqrt{(n-2)S_{xx}/SS_R}$ = 3610.047, $\hat{\alpha}$ = 0.7508, $\hat{\beta}$ = 0.001289.

H$_0$: β = 0 against H$_1$: $\beta \neq 0$.

TS: T = [$\sqrt{(n-2)S_{xx}/SS_R}$]$\times \hat{\beta}$ = (3610.047)(0.001289)
        = 4.6536.

Conclusion: Since $|T| = 4.65 > 2.179$, reject H_0. That is, there is sufficient sample evidence to claim that the number of cigarettes smoked has an effect on the death rate due to bladder cancer at the 5% level of significance.

(d) $t_{12, 0.005} = 3.055$. Same conclusion as in part (c).

7. (a) The scatter diagram of number of cigarettes smoked versus death

rate from lung cancer is shown below.

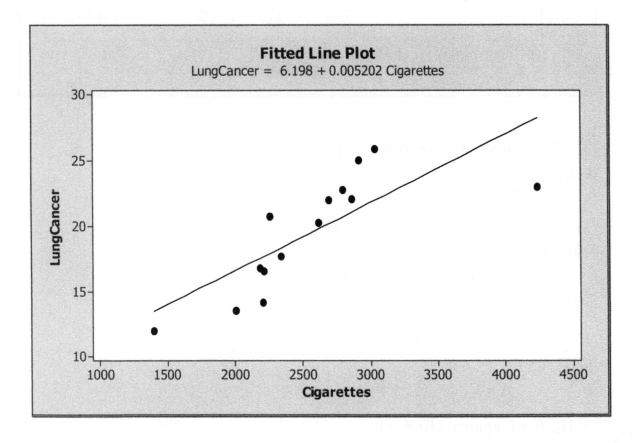

(b) The estimated regression line is $Y = 6.198 + 0.005202x$.

(c) $n = 14$, $\gamma = 0.05$, $t_{12, 0.025} = 2.179$, $S_{xx} = 5,510,467$,

$S_{YY} = 247.7415$, $S_{xY} = 28,666.88$, $SS_R = 98.60897$,

$$\sqrt{(n-2)S_{xx}/SS_R} = 818.892, \;\; \hat{\alpha} = 6.1976, \;\; \hat{\beta} = 0.005202.$$

H_0: $\beta = 0$ against H_1: $\beta \neq 0$.

TS: $T = [\sqrt{(n-2)S_{xx}/SS_R}] \times \hat{\beta} = (818.892)(0.005202)$
$= 4.260089$.

Conclusion: Since $|T| = 4.2601 > 2.179$, reject H_0. That is,
there is sufficient sample evidence to claim that the number of
cigarettes smoked has an effect on the death rate due to lung
cancer at the 5% level of significance.

(d) $t_{12, \, 0.005} = 3.055$. Same conclusion as in part (c).

9. Let Y = amount of damage.

Let x = distance.

The estimated regression line is $Y = 14.8 + 2.43x$

$n = 6$, $\gamma = 0.05$, $t_{4, \, 0.025} = 2.776$, $S_{xx} = 44.3683$,
$S_{YY} = 651.008$, $S_{xY} = 107.778$, $SS_R = 389.20$,

$$\sqrt{(n-2)S_{xx}/SS_R} = 0.6753, \;\; \hat{\alpha} = 14.8, \;\; \hat{\beta} = 2.43.$$

H_0: $\beta = 0$ against H_1: $\beta \neq 0$.

$$\text{TS: } T = [\sqrt{(n-2)S_{xx}/SS_R}\,] \times \hat{\beta} = (0.6753)(2.43) = 1.641.$$

Conclusion: Since $|T| = 1.64 < 2.776$, do not reject H_0. That is,
there is insufficient sample evidence to claim that the distance to the
nearest fire station has an effect on the fire damage sustained by a
property at the 5% level of significance.

11. (a) Let Y = amount of bananas consumed.

Let x = amount of apples consumed.

The estimated regression line is $Y = 5.0 + 0.905x$

$n = 7, \gamma = 0.05, t_{5,\,0.025} = 2.571, S_{xx} = 11.1771,$

$S_{YY} = 47.7286, S_{xY} = 10.1143, SS_R = 38.576,$

$$\sqrt{(n-2)S_{xx}/SS_R} = 1.2036, \hat{\alpha} = 5.0, \hat{\beta} = 0.905.$$

$H_0: \beta = 0$ against $H_1: \beta \neq 0$.

$$\text{TS: } T = [\sqrt{(n-2)S_{xx}/SS_R}\,] \times \hat{\beta} = (1.2036)(0.905) = 1.089.$$

Conclusion: Since $|T| = 1.089 < 2.571$, do not reject H_0. That is,
there is insufficient sample evidence to claim that the

amount of
apples consumed has an effect on the amount of bananas
consumed
at the 5% level of significance.

(b) Let Y = amount of bananas consumed.

Let x = amount of oranges consumed.

The estimated regression line is Y = 38.5 -

1.21x

$n = 7, \gamma = 0.05, t_{5, 0.025} = 2.571, S_{xx} =$
15.5143,

$S_{YY} = 47.7286, S_{xY} = -18.8429, SS_R =$
24.843,

$$\sqrt{(n-2)S_{xx}/SS_R} = 1.7671, \hat{\alpha} = 38, \hat{\beta} = -1.215.$$

$H_0: \beta = 0$ against $H_1: \beta \neq 0$.

$$\text{TS: } T = [\sqrt{(n-2)S_{xx}/SS_R}] \times \hat{\beta} = (1.7671)(-0.215) = -$$
2.147.

Conclusion: Since $|T| = 2.147 < 2.571$, do not reject H_0.
That is,
there is insufficient sample evidence to claim that the
amount of
oranges consumed has an effect on the amount of
bananas consumed at the 5% level of significance.

(c) Let Y = amount of oranges consumed.

Let x = amount of apples consumed.

The estimated regression line is Y = 11.7 +

0.148x.

$n = 7$, $\gamma = 0.05$, $t_{5, 0.025} = 2.571$, $S_{xx} = 11.1771$,

$S_{YY} = 15.5143$, $S_{xY} = 1.6586$, $SS_R = 15.268$,

$$\sqrt{(n-2)S_{xx}/SS_R} = 1.9132, \; \hat{\alpha} = 11.7, \; \hat{\beta} = 0.148.$$

H_0: $\beta = 0$ against H_1: $\beta \neq 0$.

TS: $T = [\sqrt{(n-2)S_{xx}/SS_R}] \times \hat{\beta} = (1.9132)(0.148) = 0.283.$

Conclusion: Since $|T| = 0.283 < 2.571$, do not reject H_0.
That is,
there is insufficient sample evidence to claim that the amount of
apples consumed has an effect on the amount of oranges consumed at the 5% level of significance.

SECTION 12.6 - REGRESSION TO THE MEAN

PROBLEMS

1. (a) $\hat{\alpha} = 10.5$, $\hat{\beta} = 0.325$.

(b) $n = 7$, $\gamma = 0.05$, $t_{5,\,0.025} = 2.015$, $S_{xx} = 28$,

$S_{YY} = 4.3943$, $S_{xY} = 9.1$, $SS_R = 1.4368$,

$\sqrt{(n-2)S_{xx}/SS_R} = 9.8711$, $\hat{\alpha} = 10.5$, $\hat{\beta} = 0.325$.

H_0: $\beta \geq 1$ against H_1: $\beta < 1$.

TS: $T = [\sqrt{(n-2)S_{xx}/SS_R}] \times (\hat{\beta} - \beta) =$

$(9.8711)(0.325 - 1)$

$= -6.663$.

Conclusion: Since $T = -6.663 < -2.015$, reject H_0.
   That is,

there is sufficient sample evidence to claim a
regression towards the mean diameter of the parent
seed at the 5% level of significance.

3. Responses will vary.

5. (a) Let Y = 2002 SAT scores.

Let x = 2000 SAT scores.

The regression line is $Y = -26.37 + 1.051x$.

(b) $n = 10$, $\gamma = 0.05$, $t_{8,\,0.025} = 2.306$, $S_{xx} = 11{,}564$,

$S_{YY} = 12{,}965$, $S_{xY} = 176$, $SS_R = 198$,

$$\sqrt{(n-2)S_{xx}/SS_R} = 21.6156, \ \hat{\alpha} = -26.4, \ \hat{\beta} =$$

1.05.

$H_0: \beta \geq 1$ against $H_1: \beta < 1$.

TS: $T = [\sqrt{(n-2)S_{xx}/SS_R}] \times (\hat{\beta} - \beta) =$

$(21.6156)(1.05 - 1)$

$\quad\quad = 1.0808.$

Conclusion: Since $T = 1.0808 > -2.306$, do not reject H_0. That is, there is insufficient sample evidence to claim a regression towards the mean at the 5% level of significance.

7. The distribution of the weights appears to be skewed to the right. *Note:* If the actual data values were provided one would be able to perform a normality test to establish whether one can assume a normal distribution.

SECTION 12.7 - PREDICTION INTERVALS
FOR
FUTURE RESPONSES

PROBLEMS

1. $n = 3$, $\gamma = 0.05$, $t_{1,\,0.025} = 12.706$, $S_{xx} = 8.6667$, $SS_R = 0.154$,

$\bar{x} = 2.6667$, $x_0 = 4$, $\hat{\alpha} = 2.77$, $\hat{\beta} = 2.46$.

$$W = \sqrt{\left[1 + \frac{1}{n} + \frac{(x_0 - \bar{x})^2}{S_{xx}}\right] \frac{SS_R}{(n-2)}} = 0.4867.$$

(a) When $x = 4$, the predicted $Y = 2.77 + 2.46(4) = 12.61$.

(b) The 95% prediction interval when $x = 4$ is

$12.61 \pm (12.706)(0.4867)$ or 12.61 ± 6.184.

That is, we can be 95% confident that the predicted value of Y when
$x = 4$ will lie between 6.426 and 18.794.

3. $n = 7$, $\gamma = 0.05$, $t_{5,\,0.025} = 2.571$, $\gamma = 0.1$, $t_{5,\,0.05} = 2.015$, $S_{xx} = 697.429$,

$SS_R = 8.46$, $\bar{x} = 28.286$, $x_0 = 31$, $\hat{\alpha} = 45.421$, $\hat{\beta} = -0.6563$.

$$W = \sqrt{\left[1 + \frac{1}{n} + \frac{(x_0 - \bar{x})^2}{S_{xx}}\right] \frac{SS_R}{(n-2)}} =$$

1.397.

(a) The estimated regression line is Y = 45.421 - 0.6563x.

(b) When x = 31 (salary), the predicted Y (% spent on food) is

Y = 45.421 - 0.6563(31) = 25.0757.

(c) The 95% prediction interval when x = 31 is

25.0757 ± (2.571)(1.397) or 25.0757 ± 3.5917.

That is, we can be 95% confident that the proportion of income spent on food for a family of 4 with an income of $31,000 will lie between 21.484and28.6674.

(d) The 99% prediction interval when x = 31 is

25.0757 ± (4.032)(1.397) or 25.0757 ± 5.634.

That is, we can be 99% confident that the proportion of income spent on food for a family of 4 with an income of $31,000 will lie between 19.4416and30.7099.

5. n = 12, γ = 0.05, $t_{10,\ 0.025}$ = 2.228, S_{xx} = 0.000106, SS_R = 0.000118,

$\bar{x}$ = 1.518, x_0 = 1.52, $\hat{\alpha}$ = -3.72, $\hat{\beta}$ = 4.09.

$$W = \sqrt{\left[1 + \frac{1}{n} + \frac{(x_0 - \bar{x})^2}{S_{xx}}\right] \frac{SS_R}{(n-2)}} = $$

0.0038.

(a) The estimated regression line is Y = -3.72 + 4.09x.

When x = 1.52 (refractive index), the predicted Y

(density) is

Y = -3.73 + 4.09(1.52) = 2.501.

(b) The 95% prediction interval when x = 1.52 is

3.73 ± (2.228)(0.0038) or 3.18 ± 0.008.

That is, we can be 95% confident that the density of the glass
fragment with refractive index 1.52 will lie between 2.493 and 2.510.

7. n = 8, γ = 0.05, $t_{6, 0.025}$ = 2.447, S_{xx} = 1.2289, SS_R = 31.30,

$\bar{x}$ = 3.0875, $\hat{\alpha}$ = -3.228, $\hat{\beta}$ = 12.584.

(a) The estimated regression line is Y = -3.228 + 12.584x.

When x = 2.9(grade point average in accounting),
the predicted Y (annual salary) is Y = -3.228 + 12.584(2.9)
= 33.2656. Thus, the predicted salary is $33,266.

(b) $x_0 = 2.9$, $W = \sqrt{\left[1 + \dfrac{1}{n} + \dfrac{(x_0 - \bar{x})^2}{S_{xx}}\right] \dfrac{SS_R}{(n-2)}} = 2.4532$.

The 95% prediction interval when x = 2.9 is

$33.2656 \pm (2.447)(2.4532)$ or 33.2656 ± 6.003.

That is, we can be 95% confident that the annual salary with a grade point average of 2.9 in accounting will lie between 27.2626 ($27,263) and 39.2686 ($39,267).

(c) $x_0 = 3.6$, $W = \sqrt{\left[1 + \dfrac{1}{n} + \dfrac{(x_0 - \bar{x})^2}{S_{xx}}\right] \dfrac{SS_R}{(n-2)}} = 2.6427$.

When x = 3.6(grade point average in accounting), the predicted Y (annual salary) is Y = -3.228 + 12.584(3.6) = 42.0744. Thus, the predicted salary is $42,074.

The 95% prediction interval when x = 3.6 is

$42.0744 \pm (2.447)(2.6427)$ or 42.0744 ± 6.4667.

That is, we can be 95% confident that the annual salary with a grade point average of 3.6 in accounting will lie between 35.6077 ($35,608) and 48.5411 ($48,541).

SECTION 12.8 - COEFFICIENT OF DETERMINATION

PROBLEMS

1. (a) The scatter diagram, along with the regression line, of
   selling price versus square footage are shown below.

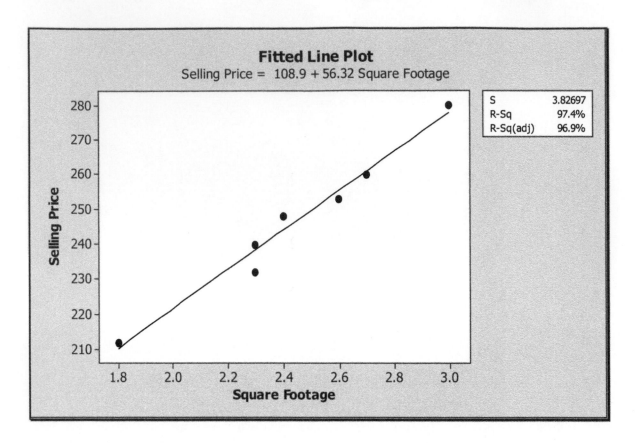

Fitted Line Plot
Selling Price = 108.9 + 56.32 Square Footage

S	3.82697
R-Sq	97.4%
R-Sq(adj)	96.9%

(b) The estimated regression line is Y = 108.9 + 56.32x.

(c) $S_{YY} = 2791.71$, $SS_R = 73.23$.

The coefficient of determination:

$$R^2 = \frac{S_{YY} - SS_R}{S_{YY}} = (2791.71 - 73.2)/2791.71 = 0.9738.$$

Thus, 97.38% of the variability of the selling price is explained by the square footage.

Note: The coefficient of determination is also given in the output as 97.4%.

(d) $n = 7$, $\gamma = 0.05$, $t_{5, 0.025} = 2.571$, $S_{xx} = 0.8571$, $SS_R = 73.2$,

$\bar{x} = 2.4429$, $x_0 = 2.6$.

$$W = \sqrt{\left[1 + \frac{1}{n} + \frac{(x_0 - \bar{x})^2}{S_{xx}}\right] \frac{SS_R}{(n-2)}} = 4.1416.$$

The predicted selling price when $x = 2.6$ (i.e. 2600 square feet) is

$Y = 108.9 + 56.32(2.6) = 255.332$.

The 95% prediction interval when $x = 2.6$ is

$255.332 \pm (2.571)(4.1416)$ or 255.332 ± 10.6481.

That is, we can be 95% confident that the selling price of a three-bedroom home with a size of 2.6 (2600 square feet) will lie between 244.6839 ($244,683.9) and 265.9801 ($265,980.1).

3. (a) $n = 5$, $S_{YY} = 1860.8$, $SS_R = 0.697198$.

   The coefficient of determination:

   $$R^2 = \frac{S_{YY} - SS_R}{S_{YY}} = (1860.8 - 0.6971)/1860.8 = 0.9996.$$

   (b) This is a reasonable way of estimating the amount of protein in a liver sample. From part (a), 99.96% of the variability in the amount of protein is explained by the absorbed light.

   (c) The estimated regression line is $Y = -15.0951 + 38.04669x$. So when $x = 1.5$, the estimate of the amount of protein is $Y = -15.0951 + 38.04669(1.5) = 41.9749$.

   (d) $n = 5$, $\gamma = 0.1$, $t_{3, 0.05} = 2.353$, $S_{xx} = 1.285$, $SS_R = 0.697198$,

$\bar{x} = 1.18$, $x_0 = 1.5$.

$$W = \sqrt{\left[1 + \frac{1}{n} + \frac{(x_0 - \bar{x})^2}{S_{xx}}\right] \frac{SS_R}{(n-2)}} = 0.545343.$$

The 90% prediction interval when $x = 1.5$ is

$41.9749 \pm (2.353)(0.5453)$ or 41.9749 ± 1.28.

That is, we can be 90% confident that the amount of protein when the light absorbed is 1.5 will lie between 40.6918 and 43.2582.

5. $S_{YY} = 872.25$, $SS_R = 742.21$.

The coefficient of determination:

$$R^2 = \frac{S_{YY} - SS_R}{S_{YY}} = (872.25 - 742.21)/872.25 = 0.1491.$$

7. $S_{YY} = 53.3$, $SS_R = 50.1197$.

The coefficient of determination:

$$R^2 = \frac{S_{YY} - SS_R}{S_{YY}} = (53.3 - 50.1197)/53.3 = 0.0597.$$

SECTION 12.9 - SAMPLE CORRELATION COEFFICIENT

PROBLEMS

1. (a) $S_{YY} = 14$, $SS_R = 0.286$, $S_{xx} = 4.667$, $S_{xY} = 8$.

The coefficient of determination:

$$R^2 = \frac{S_{YY} - SS_R}{S_{YY}} = 0.9796.$$

The correlation coefficient is $+ \sqrt{0.9796} = + 0.9897$. The correlation is positive since S_{xY} is positive. Alternatively, you can use $r = \dfrac{S_{xY}}{\sqrt{S_{xx} \cdot S_{YY}}}$

(b) The result is the same as in part (a). Observe that the values are just switched around. Thus, the correlation between x and Y is same as the correlation between Y and x.

3. (a) Since the slope is positive, $r = + \sqrt{0.64} = +0.8$.

(b) Since the slope is positive, $r = + \sqrt{0.64} = +0.8$.

(c) Since the slope is negative, $r = - \sqrt{0.64} = -0.8$.

(d) Since the slope is negative, $r = - \sqrt{0.64} = -0.8$.

5. Positive correlation.

(a) Let Y = husband's age.

Let x = wife's age.

The estimated regression line is Y = -3.163 + 1.241x.

(b) Let Y = wife's age.

Let x = husband's age.

The estimated regression line is Y = 7.248 + 0.6589x.

(c) The coefficient of determination, R^2 = 0.818. Since the slope for the regression line in part (a) is positive, then the correlation coefficient will also be positive. Thus,
r = +0.9044.

(d) The coefficient of determination, R^2 = 0.818 (same as in (c)).
Since the slope for the regression line in part (a) is positive, then the correlation coefficient r = +0.9044.

SECTION 12.10 – ANALYSIS OF RESIDUALS:
ASSESSING THE MODEL

PROBLEMS

1. Below are two graphs. The first shows the regression line superimposed on the scatter plot. The second, shows the

residual plot of the standardized residuals versus the independent variable of training time. From the random nature of the scatter diagram and the residual plot, it seem that the linear model is appropriate. Also, observe that the residual plot shows a pattern, in that the absolute value of the residuals, and thus their squares, appears to be increasing, as the x values increase. This would indicate that the constant variance assumption for the response variable has been violated and that the variance seem to be increasing.

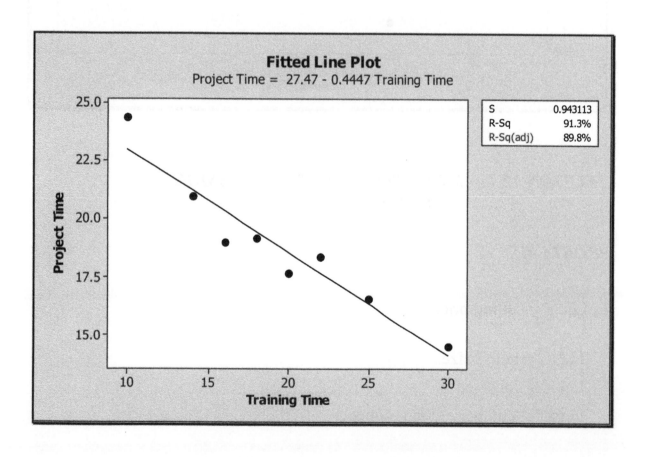

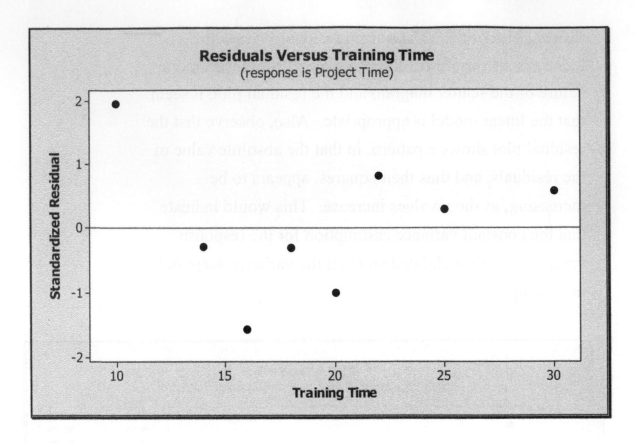

Residuals Versus Training Time
(response is Project Time)

SECTION 12.11 - MULTIPLE LINEAR REGRESSION MODEL

PROBLEMS

1. Let Y = selling price.

   Let x_1 = house size.

   Let x_2 = lot size.

   Let x_3 = number of bath rooms.

   Below is a partial *Minitab* output for this problem.

```
Regression Analysis: Y versus X1, X2, X3

The regression equation is
Y = 101 + 0.0329 X1 + 43.6 X2 + 10.4 X3
```

```
Predictor         Coef    SE Coef       T       P
Constant       100.985      7.862   12.84   0.000
X1            0.032945   0.009069    3.63   0.011
X2               43.65      29.38    1.49   0.188
X3              10.394      6.863    1.51   0.181


S = 5.26997    R-Sq = 97.4%    R-Sq(adj) = 96.0%


Analysis of Variance

Source            DF       SS       MS       F       P
Regression         3   6159.0   2053.0   73.92   0.000
Residual Error     6    166.6     27.8
Total              9   6325.6


Source   DF   Seq SS
X1        1   5969.0
X2        1    126.3
X3        1     63.7
```

(a) The estimated multiple regression equation is

$$Y = 101 + 0.0329x_1 + 43.6x_2 + 10.4x_3.$$

Note: The coefficients are given to the first decimal place.

(b) For $x_1 = 2500$, $x_2 = 0.4$, and $x_3 = 2$
$\Rightarrow Y = 221.49$ ($\$221,490$).

(c) For $x_1 = 2500$, $x_2 = 0.4$, and $x_3 = 3$
$\Rightarrow Y = 231.89$ ($\$231,890$).

3. The estimated multiple regression equation is

$$Y = -153.513 + 51.744x_1 + 0.077x_2 + 20.923x_3 + 13.103x_4.$$

For $x_1 = 2$, $x_2 = 7$, $x_3 = 3$, $x_4 = 13 \Rightarrow Y = 183.622$.

5. Let Y = hardness.

    Let x_1 = copper content.

    Let x_2 = annealing temperature.

    Below is a partial *Minitab* output for this problem.

```
Regression Analysis: Y versus X1, X2

The regression equation is
Y = 154 + 19.3 X1 - 0.0754 X2


Predictor        Coef     SE Coef        T       P
Constant       153.84       10.43    14.74   0.000
X1              19.30       23.12     0.84   0.431
X2          -0.075431    0.008122    -9.29   0.000


S = 3.52168    R-Sq = 92.8%    R-Sq(adj) = 90.8%


Analysis of Variance

Source            DF        SS       MS       F       P
Regression         2   1122.45   561.22   45.25   0.000
Residual Error     7     86.82    12.40
Total              9   1209.26


Source   DF   Seq SS
X1        1     52.73
X2        1   1069.72
```

The estimated multiple regression equation is

$Y = 154 + 19.3x_1 - 0.0754x_2$.

For $x_1 = 0.15$, $x_2 = 1150$

$\Rightarrow Y = 70.485$. Thus, the estimated mean hardness of the steel is 70.485 (units).

REVIEW PROBLEMS

1. (a) The scatter diagram for this problem is shown below.

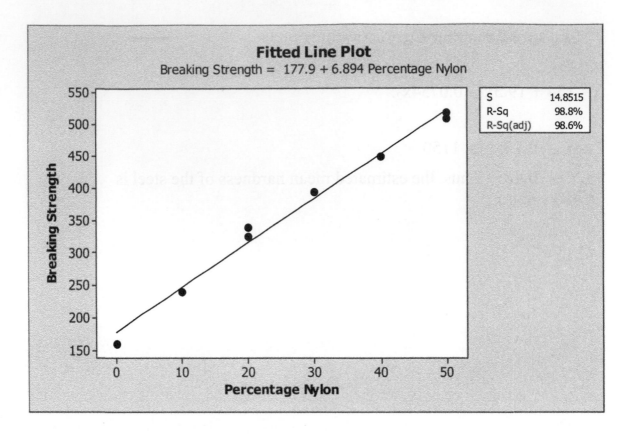

Fitted Line Plot
Breaking Strength = 177.9 + 6.894 Percentage Nylon

S	14.8515
R-Sq	98.8%
R-Sq(adj)	98.6%

(b) Let Y = breaking strength.

Let x = percentage nylon.

The estimated regression line is $Y = 177.9 + 6.894x$.

(c) The predicted breaking strength when x = 50 is
$Y = 522.626$.

(d) $n = 8$, $\gamma = 0.05$, $t_{6,\,0.025} = 2.447$, $S_{xx} = 2350$,
$SS_R = 1323$, $\bar{x} = 27.5$, $x_0 = 50$.

$$W = \sqrt{\left[1 + \frac{1}{n} + \frac{(x_0 - \bar{x})^2}{S_{xx}}\right]\frac{SS_R}{(n-2)}} = 17.192.$$

The 95% prediction interval when x = 50 is

$522.626 \pm (2.447)(17.192)$ or 522.626 ± 42.069.

That is, we can be 95% confident that the breaking strength of the rope when the percentage of nylon is 50 will lie between 480.557 and 564.695.

3. (a) Let Y = systolic blood pressure of the females.

   Let x = weight of the females.

   The estimated regression line is $Y = 94.3 + 0.155x$.

   (b) $n = 20$, $\gamma = 0.05$, $t_{18, 0.025} = 2.101$, $S_{xx} = 5100.55$,
   $SS_R = 1482.17$, $\bar{x} = 125.15$, $x_0 = 120$.

   $$W = \sqrt{\left[1 + \frac{1}{n} + \frac{(x_0 - \bar{x})^2}{S_{xx}}\right] \frac{SS_R}{(n-2)}} = 9.321.$$

   When $x_0 = 120$, $Y = 112.9$.

   The 95% prediction interval when $x = 120$ is

   $112.9 \pm (2.101)(9.321)$ or 112.9 ± 19.583.

   That is, we can be 95% confident that the systolic blood pressure of a female when her weight is 120 pounds will lie between 93.317and 132.483.

   (c) There are a total of eight females with weights between 119 and 121 pounds. Out of the eight students, all eight (100%) had systolic blood pressure between 93.317 and 132.483.

5. We cannot conclude that verbal praise tends to lower performance

levels whereas verbal criticism tends to raise them because one does not necessarily causes the other. However there is an association (correlation) between the variables.

7. Let the base year of 1990 be equivalent to zero, 1991 be equivalent to 1, etc.

(a) Y = gasoline retail prices per gallon in the U.S.

x = year.

The estimated regression line is $Y = 1.069 + 0.02462x$.

(b) $n = 13$, $\gamma = 0.05$, $t_{11, 0.025} = 2.201$, $S_{xx} = 182$, $SS_R = 0.1264$,

$$\sqrt{(n-2)S_{xx}/SS_R} = 125.8515, \quad \hat{\alpha} = 1.069, \quad \hat{\beta} = 0.02462.$$

$H_0: \beta = 0$ against $H_1: \beta \neq 0$.

TS: $T = [\sqrt{(n-2)S_{xx}/SS_R}] \times \hat{\beta} = (125.8515)(0.02462) = 3.0985$.

Conclusion: Since $|T| = 3.0985 > 2.201$, reject H_0. That is, there is sufficient sample evidence to claim that the slope of the regression line is not equal to zero at the 5% level of significance. That is, there is sufficient sample evidence to claim that the year had an effect on the price per gallon of gasoline U.S.

9. Let Y = percentage of manufacturing workers who were union members in 1989.

Let x = percentage of manufacturing workers who were union members in 1984.

(a) The predicted equation is $Y = -1.374 + 0.871x$.

Thus, the predicted 1989 percentage of Ohio's manufacturing workers who were union members is
$Y = -1.374 + 0.871(41.6) = 34.86$.

(b) $n = 9$, $\gamma = 0.05$, $t_{7, 0.025} = 2.365$, $S_{xx} = 1129.39$,
$SS_R = 89.64$, $\bar{x} = 28.289$, $x_0 = 17.5$.

$$W = \sqrt{\left[1 + \frac{1}{n} + \frac{(x_0 - \bar{x})^2}{S_{xx}}\right] \frac{SS_R}{(n-2)}} = 3.943.$$

When $x_0 = 17.5$, $Y = 13.869$.
The 95% prediction interval when $x = 17.5$ is

$13.869 \pm (2.365)(3.943)$ or 13.869 ± 9.325.

That is, we can be 95% confident that Oklahoma's 1989 membership percentage when its 1984 percentage membership was 17.5 will lie between 4.544 and 132.483.

11. Let Y = tensile strength of the synthetic fiber.

Let x_1 = percentage of cotton in the fiber.

Let x_2 = drying time of the fiber.

(a) A partial *Minitab* output for this problem is shown below. The multiple regression equation with the tensile strength as the response variable is $Y = 177.41 + 1.07x_1 + 11.701x_2$.

```
Regression Analysis: Y versus X1, X2

The regression equation is
Y = 177 + 1.07 X1 + 11.7 X2


Predictor      Coef   SE Coef      T       P
Constant     177.41     19.42   9.14   0.000
X1            1.070     1.184   0.90   0.407
X2           11.701     3.314   3.53   0.017


S = 6.94729    R-Sq = 78.4%    R-Sq(adj) = 69.7%


Analysis of Variance

Source          DF        SS       MS      F      P
Regression       2    874.18   437.09   9.06  0.022
Residual Error   5    241.32    48.26
Total            7   1115.50


Source   DF   Seq SS
X1        1   272.55
X2        1   601.62
```

(b) When $x_1 = 22$ and $x_2 = 3.5$, then the predicted tensile strength is
$Y = 177.41 + 1.07(22) + 11.701(3.5) = 241.904$.

13. Let Y = job satisfaction.

Let x_1 = yearly income (in $1,000).

Let x_2 = years on the job.

(a) A partial *Minitab* output for this problem is shown below.
The multiple regression equation with job satisfaction as the
response variable is $Y = -0.40 + 0.162x_1 - 0.113x_2$.

```
Regression Analysis: Y versus X1, X2

The regression equation is
Y = - 0.40 + 0.162 X1 - 0.113 X2


Predictor       Coef   SE Coef      T      P
Constant      -0.395     2.079  -0.19  0.855
X1           0.16195   0.05512   2.94  0.026
X2          -0.11283   0.08001  -1.41  0.208


S = 0.369199   R-Sq = 82.6%   R-Sq(adj) = 76.8%


Analysis of Variance

Source            DF      SS      MS      F      P
Regression         2  3.8910  1.9455  14.27  0.005
Residual Error     6  0.8178  0.1363
Total              8  4.7089


Source  DF  Seq SS
X1       1  3.6200
X2       1  0.2711
```

(b) The estimated regression equation for the job satisfaction ratings is $Y = -0.40 + 0.162x_1 - 0.113x_2$. Here, when the yearly income remains fixed, the job satisfaction rating decreases by 0.113 when the number of years on the job increases by one year.

(c) When $x_1 = 51$ ($51,000), $x_2 = 5$, the predicted job rating is

$$Y = -0.40 + 0.162(51) - 0.113(5) = 7.297.$$

15. Rejecting the hypothesis that cigarette smoking and bladder cancer rates are unrelated does not imply that cigarette

smoking directly leads to getting bladder cancer. This may
be caused by constant alcohol consumption or even
your age. In order to determine what other variables may
cause bladder cancer, one can perform a multiple regression
analysis on the bladder cancer rates and the other
independent variables of interest. The analysis will indicate
their association.

Chapter 13 CHI-SQUAREDGOODNESS-OF-FIT TESTS

SECTION 13.2 - CHI-SQUARED GOODNESS-OF-FIT TESTS

PROBLEMS

1. (a) $\chi^2_{5,0.01} = 15.09$.   (b) $\chi^2_{5,0.05} = 11.07$.   (c) $\chi^2_{10,0.01} = 23.21$

   (d) $\chi^2_{10,0.05} = 18.31$.   (e) $\chi^2_{20,0.05} = 31.41$.

3. $n = 300$, $N_1 = 148$, $N_2 = 92$, $N_3 = 60$, $P_1 = 0.52$, $P_2 = 0.32$,

   $P_3 = 0.16$, $e_1 = (300)(0.52) = 156$, $e_2 = (300)(0.32) = 96$,

   $e_3 = (300)(0.16) = 48$, $\alpha = 0.05$, $\chi^2_{2,0.05} = 5.99$.

   H_0: $P_1 = 0.52$, $P_2 = 0.32$, $P_3 = 0.16$ vs. H_1: At least one P_i is different.

   TS: $\chi^2 = 3.5769$.

   Conclusion: Since $3.5769 < 5.99$, do not reject H_0. That is, there is insufficient sample evidence to claim that the percentages are different at the 5% level of significance.

5. $n = 60$, $N_1 = 4$, $N_2 = 3$, $N_3 = 7$, $N_4 = 17$, $N_5 = 16$, $N_6 = 13$,

   $P_1 = 0.1$, $P_2 = 0.1$, $P_3 = 0.05$, $P_4 = 0.4$, $P_5 = 0.2$, $P_6 = 0.15$

   $e_1 = (60)(0.1) = 6$, $e_2 = (60)(0.1) = 6$, $e_3 = (60)(0.05) = 3$,

   $e_4 = (60)(0.4) = 24$, $e_5 = (60)(0.2) = 12$, $e_6 = (60)(0.15) = 9$,

$\alpha = 0.05$, $\chi^2_{5,0.05} = 11.07$. *Note:* $e_3 < 5$.

H_0: $P_1 = 0.1$, $P_2 = 0.1$, $P_3 = 0.05$, $P_4 = 0.4$, $P_5 = 0.2$, $P_6 = 0.15$

H_1: At least one P_i is different.

TS: $\chi^2 = 12.6528$.

Conclusion: Since $12.6528 > 11.07$, reject H_0. That is, there is sufficient sample evidence to claim that at least one of the probabilities is different at the 5% level of significance.

7. $n = 500$, $N_1 = 222$, $N_2 = 171$, $N_3 = 98$, $N_4 = 9$,

   $P_1 = 0.38$, $P_2 = 0.32$, $P_3 = 0.26$, $P_4 = 0.04$,

   $e_1 = (500)(0.38) = 190$, $e_2 = (500)(0.32) = 160$, $e_3 = (50)(0.26) = 130$,

   $e_4 = (500)(0.04) = 20 = 9$,

   H_0: $P_1 = 0.38$, $P_2 = 0.32$, $P_3 = 0.26$, $P_4 = 0.04$

   H_1: At least one P_i is different.

   TS: $\chi^2 = 20.0726$.

   p-value $= P[\chi^2 > 20.0726] < 0.005$ (with 3 degrees of freedom).

   Conclusion: Since the p-value is rather small, reject H_0. That is, there is sufficient sample evidence to claim that at least one of the probabilities is different.

9. $n = 400$, $N_1 = 106$, $N_2 = 138$, $N_3 = 76$, $N_4 = 80$,

   $P_i = 0.25$ for $i = 1, 2, 3, 4$,

   $e_i = (400)(0.25) = 100$, for $i = 1, 2, 3, 4$

$\alpha = 0.05$, $\chi^2_{3,0.05} = 7.81$.

H_0: $P_i = 0.25$, for $i = 1, 2, 3, 4$.

H_1: At least one P_i is different.

TS: $\chi^2 = 24.56$.

Conclusion: Since $24.56 > 7.81$, reject H_0. That is, there is sufficient sample evidence to claim that at least one of the probabilities is different at the 5% level of significance. The employee is correct at the 5% level of significance.

For $\alpha = 0.01$, $\chi^2_{3,0.01} = 11.34$, the decision is the same.

11. $n = 10{,}000$, $N_1 = 1122$, $N_2 = 1025$, $N_3 = 1247$, $N_4 = 818$, $N_5 = 1043$, $N_6 = 827$, $N_7 = 1147$, $N_8 = 946$, $N_9 = 801$, $N_{10} = 1022$,

$P_1 = 0.1$, for $i = 1, 2, 3, \ldots, 10$.

$e_1 = (10{,}000)(0.1) = 1{,}000$ for $i = 1, 2, 3, \ldots, 10$.

H_0: $P_1 = 0.1$ for $I = 1, 2, 3, \ldots, 10$.

H_1: At least one P_i is different.

TS: $\chi^2 = 206.03$.

p-value $= P[\chi^2 > 206.03] <<< 0.005$ (with 9 degrees of freedom).

Conclusion: Since the p-value is very small, reject H_0. That is, there is sufficient sample evidence to claim that at least one of the probabilities is different. That is, the numbers are not being played with equal frequency.

13. $n = 400$, $N_1 = 225$, $N_2 = 112$, $N_3 = 33$,

$P_1 = 0.598$, $P_2 = 0.324$, $P_3 = 0.078$,

$e_1 = (400)(0.598) = 239.2$, $e_2 = (400)(0.324) = 129.6$,

$e_3 = (400)(0.078) = 31.2$, $\alpha = 0.05$, $\chi^2_{2,0.05} = 5.99$.

H_0: $P_1 = 0.598$, $P_2 = 0.324$, $P_3 = 0.078$

H_1: At least one P_i is different.

TS: $\chi^2 = 3.5376$.

Conclusion: Since $3.5376 < 5.99$, do not reject H_0. That is, there is insufficient sample evidence to claim that the percentages are different in 2002 at the 5% level of significance.

15. $n = 1000$, $N_1 = 42$, $N_2 = 403$, $N_3 = 315$, $N_4 = 240$,

$P_1 = 0.011$, $P_2 = 0.32$, $P_3 = 0.36$, $P_4 = 0.309$ (sum of the last two)

$e_1 = (1000)(0.011) = 11$, $e_2 = (1000)(0.32) = 320$,

$e_3 = (1000)(0.36) = 360$, $e_4 = (1000)(0.309) = 309$.

H_0: $P_1 = 0.011$, $P_2 = 0.32$, $P_3 = 0.36$, $P_4 = 0.309$.

H_1: At least one P_i is different.

TS: $\chi^2 = 24.56$.

p-value $= P[\chi^2 > 129.9245] <<< 0.005$ (with 3 degrees of freedom).

Conclusion: Since the p-value is very small, reject H_0. That is, there is sufficient sample evidence to claim that at least one of the probabilities is different. That is, these observed percentages are not the same as the 1986 percentages.

SECTION 13.3 TESTING FOR INDEPENDENCE IN POPULATIONS CLASSIFIED ACCORDING TO TWO CHARACTERISTICS

PROBLEMS

1. Below is a *Minitab* output with the relevant computations for this problem.

```
Chi-Square Test: A, B, C

Expected counts are printed below observed counts
Chi-Square contributions are printed below expected counts

           A      B      C   Total
   1      32     12     40      84
        29.81  20.32  33.87
        0.161  3.408  1.109

   2      56     48     60     164
        58.19  39.68  66.13
        0.083  1.746  0.568

Total     88     60    100     248

Chi-Sq = 7.075, DF = 2, P-Value = 0.029
```

(a) TS: $\chi^2 = 7.075$ with a p-value = 0.029.

(b) H_0: the X characteristic and the Y characteristic are independent.

H_1: the X characteristic and the Y characteristic are not independent.

TS: $\chi^2 = 7.075$ with a p-value = 0.029.

Conclusion: Since p-value = 0.029 < α = 0.05, reject H_0. That is, there is sufficient sample evidence to conclude that the X characteristic and the Y characteristic are not independent.

(c) At the 1% level of significance, p-value = 0.029 > α = 0.01, do not reject H_0. That is, there is insufficient sample evidence to conclude that the X characteristic and the Y characteristic are not independent.

3. Below is a *Minitab* output with the relevant computations for this problem.

```
Chi-Square Test: Women, Men

Expected counts are printed below observed counts
Chi-Square contributions are printed below expected counts

                            Women      Men       Total
    Positive Evaluation       54        47         101
                            52.39     48.61
                            0.049     0.053

    Negative Evaluation       20        32          52
                            26.97     25.03
                            1.803     1.943

    Not Sure                  23        11          34
                            17.64     16.36
                            1.631     1.758

            Total            97        90         187

Chi-Sq = 7.238, DF = 2, P-Value = 0.027
```

H_0: The evaluation of the U.S. President and the gender of the evaluator are independent.

H_1: The evaluation of the U.S. President and the gender of the evaluator are not independent.

TS: χ^2 = 7.238 with a p-value = 0.027.

Conclusion: Since p-value $= 0.027 < \alpha = 0.05$, reject H_0. That is, there is sufficient sample evidence at the 5% significance level, to conclude that the evaluation of the President and the gender of the evaluator are not independent.

5. Below is a *Minitab* output with the relevant computations for this problem.

```
Chi-Square Test: Economy, Standard, Luxury

Expected counts are printed below observed counts
Chi-Square contributions are printed below expected counts

                Economy  Standard  Luxury  Total
    Excellent        30        21       9     60
                  30.19     22.45    7.35
                  0.001     0.094   0.368

    Good             36        29       8     73
                  36.74     27.32    8.95
                  0.015     0.104   0.101

    Fair             12         8       2     22
                  11.07      8.23    2.70
                  0.078     0.007   0.180

Total                78        58      19    155

Chi-Sq = 0.947, DF = 4, P-Value = 0.918
1 cells with expected counts less than 5.
```

H_0: The service of the hotel guests and the price of the room are independent.

H_1: The service of the hotel guests and the price of the room are not independent.

TS: $\chi^2 = 0.947$ with a p-value $= 0.918$.

Conclusion: Since p-value $= 0.918$ is rather large, do not reject H_0. That is, there is insufficient sample evidence to conclude that the service of the hotel guests and the price of the room are not independent.

7. Below is a *Minitab* output with the relevant computations for this problem.

```
┌─────────────────────────────────────────────────────────────────┐
│                                                                   │
│  Chi-Square Test: Two Courses, Three or More Courses              │
│                                                                   │
│  Expected counts are printed below observed counts                │
│  Chi-Square contributions are printed below expected counts       │
│                                                                   │
│                                        Three                       │
│                          Two           or More                     │
│                          Courses       Courses     Total           │
│         Above Average        10              4        14           │
│                            6.73           7.27                      │
│                           1.590          1.472                      │
│                                                                    │
│         Average              40             38        78           │
│                           37.49          40.51                      │
│                           0.168          0.156                      │
│                                                                    │
│         Below Average        12             25        37           │
│                           17.78          19.22                      │
│                           1.881          1.740                      │
│                                                                    │
│                  Total       62             67       129           │
│                                                                    │
│  Chi-Sq = 7.007, DF = 2, P-Value = 0.030                          │
│                                                                   │
└─────────────────────────────────────────────────────────────────┘
```

H_0: Teaching performance and the number of courses taught are independent.

H_1: Teaching performance and the number of courses taught are not independent.

TS: $\chi^2 = 7.007$ with a p-value = 0.03.

Conclusion: Since p-value = 0.03 < α = 0.05, reject H_0. That is, there is sufficient sample evidence to conclude that teaching performance and the number of courses taught are not independent.

9. Below is a Minitab output with the relevant computations for this problem.

```
Chi-Square Test: AgeGroup1, AgeGroup2, AgeGroup3

Expected counts are printed below observed counts
Chi-Square contributions are printed below expected counts

                  AgeGroup1  AgeGroup2  AgeGroup3  Total
    Excellent           18         20         41     79
                     20.31      20.64      38.05
                     0.264      0.020      0.229

    Good                25         27         43     95
                     24.43      24.82      45.76
                     0.013      0.192      0.166

    Fair                17         15         26     58
                     14.91      15.15      27.93
                     0.292      0.002      0.134

    Poor                 3          2          8     13
                      3.34       3.40       6.26
                     0.035      0.574      0.483

        Total          63         64        118    245

Chi-Sq = 2.403, DF = 6, P-Value = 0.879
2 cells with expected counts less than 5.
```

H_0: Age group and the shampoo ratings are independent.

H_1: Age group and the shampoo ratings are not independent.

TS: $\chi^2 = 2.403$ with a p-value = 0.879.

Conclusion: Since p-value = 0.879 > α = 0.05, do not reject H_0. That is, there is insufficient sample evidence to conclude that the age group of the individuals and the shampoo ratings are not independent.

11. Below is a *Minitab* output with the relevant computations for this problem.

```
Chi-Square Test: Doctors, Lawyers, Engineers

Expected counts are printed below observed counts
Chi-Square contributions are printed below expected counts

                 Doctors    Lawyers   Engineers     Total
   Protestant         64        110         152       326
                   92.49     110.88      122.63
                   8.774      0.007       7.032

   Catholic           60         86          78       224
                   63.55      76.19       84.26
                   0.198      1.264       0.466

   Jewish             57         21          10        88
                   24.97      29.93      3 3.10
                  41.105      2.665      16.124

        Total        181        217         240       638

Chi-Sq = 77.635, DF = 4, P-Value = 0.000
```

H_0: Profession of individuals in the organization and their religious affiliation are independent.

H_1: Profession of individuals in the organization and their religious affiliation are not independent.

TS: $\chi^2 = 77.635$ with a p-value = 0.000.

Conclusion: Since p-value = 0.000 < α = 0.05, reject H_0.
That is, there is sufficient sample evidence to conclude that the profession of individuals in the organization and their religious affiliation are not independent.

Same conclusion at the 1% significance level.

13. Below is a *Minitab* output with the relevant computations for this problem.

```
Chi-Square Test: Doctors, Lawyers, Engineers

Expected counts are printed below observed counts
Chi-Square contributions are printed below expected counts

                Doctors    Lawyers    Engineers    Total
Protestant          128        220          304      652
                 184.97     221.76       245.27
                 17.547      0.014       14.065

Catholic            120        172          156      448
                 127.10     152.38       168.53
                  0.396      2.527        0.931

Jewish              114         42           20      176
                  49.93      59.86        66.21
                 82.210      5.330       32.249

      Total         362        434          480     1276

Chi-Sq = 155.269, DF = 4, P-Value = 0.000
```

H_0: Profession of individuals in the organization and their religious affiliation are independent.

H_1: Profession of individuals in the organization and their religious affiliation are not independent.

TS: $\chi^2 = 155.269$ with a p-value = 0.000.

Conclusion: Since p-value = 0.000 < α = 0.05, reject H_0.
That is, there is sufficient sample evidence to conclude that the profession of individuals in the organization and their religious affiliation are not independent.

Same conclusion at the 1% significance level.

SECTION 13.4 - TESTING FOR INDEPENDENCE IN CONTINGENCY TABLES WITH FIXED MARGINAL TOTALS

PROBLEMS

1. No. We can say that there is an association between lung cancer and cigarette smoking.

3. Below is a *Minitab* output with the relevant computations for this problem.

```
Chi-Square Test: Accident, No Accident

Expected counts are printed below observed counts
Chi-Square contributions are printed below expected counts

                    Accident    No Accident    Total
  Cellular Phone        22            278        300
                      20.57         279.43
                      0.099           0.007

  No Phone              26            374        400
                      27.43         372.57
                      0.074           0.005

          Total         48            652        700

Chi-Sq = 0.186, DF = 1, P-Value = 0.666
```

H_0: Having a cellular phone in your car and being involved in an accident are independent.

H_1: Having a cellular phone in your car and being involved in an accident are not independent.

TS: $\chi^2 = 0.186$ with a p-value = 0.666.

Conclusion: Since p-value = 0.666 > α = 0.05, do not reject H_0. That is, there is insufficient sample evidence to conclude that having a cellular phone in your car and being involved in an accident are not independent.

5. Below is a *Minitab* output with the relevant computations for this problem.

```
Chi-Square Test: Defective, Nondefective

Expected counts are printed below observed counts
Chi-Square contributions are printed below expected counts

                Defective      Nondefective        Total
    Before          22              404              426
                 19.68           406.32
                 0.274            0.013

    After           18              422              440
                 20.32           419.68
                 0.266            0.013

    Total           40              826              866

Chi-Sq = 0.566, DF = 1, P-Value = 0.452
```

H_0: The modifications and percentage of defective items are independent.

H_1: The modifications and percentage of defective items are not independent.

TS: χ^2 = 0.566 with a p-value = 0.452.

Conclusion: Since p-value = 0.452 is rather large, do not reject H_0. That is, there is insufficient sample evidence to conclude that the modifications and percentage of defective items are not independent. That is, the modification does not result in a different percentage of defective items.

7. Below is a *Minitab* output with the relevant computations for this problem.

```
Chi-Square Test: Fluoridated, Nonfluoridated

Expected counts are printed below observed counts
Chi-Square contributions are printed below expected counts

                 Fluoridated    Nonfluoridated     Total
      0               154               133           287
                   143.50            143.50
                    0.768             0.768

      1                20                18            38
                    19.00             19.00
                    0.053             0.053

      2                14                21            35
                    17.50             17.50
                    0.700             0.700

3 or more             12                28            40
                    20.00             20.00
                    3.200             3.200

Total                200               200           400

Chi-Sq = 9.442, DF = 3, P-Value = 0.024
```

H_0: The number of dental cavities a person has and whether that person's water supply is fluoridated are independent.

H_1: The number of dental cavities a person has and whether that person's water supply is fluoridated are not independent.

TS: $\chi^2 = 9.442$ with a p-value - 0.024.

Conclusion: Since p-value = 0.024 < α = 0.05, reject H_0. That is, there is sufficient sample evidence to conclude that the number of dental cavities a person has and whether that person's water supply is fluoridated are not independent at the 5% significance level.

Do not reject H_0 at the 1% level of significance. That is, there is insufficient sample evidence to conclude that the number of dental cavities a person has and whether that person's water supply is fluoridated are not independent at the 1% significance level.

9. (a) Below is a *Minitab* output with the relevant computations for this problem. Note the p-value =0.594 (large).

For $\alpha = 0.1$, $\chi^2_{2,\,0.1} = 2.71$.

```
Chi-Square Test: Smokers, Nonsmokers

Expected counts are printed below observed counts
Chi-Square contributions are printed below expected counts

                        Smokers      Nonsmokers      Total
    College Students       18             32            50
                        18.00          32.00
                        0.000          0.000

    College Faculty        12             28            40
                        14.40          25.60
                        0.400          0.225

    Bankers                24             36            60
                        21.60          38.40
                        0.267          0.150

            Total          54             96           150

Chi-Sq = 1.042, DF = 2, P-Value = 0.594
```

H_0: $P_1 = P_2 = P_3 = 1/3$.

H_1: At least one P_i is different.

TS: $\chi^2 = 1.042$.

Conclusion: Since $1.042 < 2.71$, do not reject H_0. That is, there is insufficient sample evidence to claim that the percentages of smokers in the three groups are different at the 10% level of significance.

(b) For $\chi^2_{2,0.05} = 5.99$. Since $1.042 < 5.99$, do not reject H_0. Same conclusion as in part (a).

(c) $\chi^2_{2,0.01} = 9.21$. Same conclusion as in parts (a) and (b).

REVIEW PROBLEMS

1. $n = 219$, $N_1 = 47$, $N_2 = 52$, $N_3 = 57$, $N_4 = 63$, $P_1 = P_2 = P_3 = P_4 = 0.25$,

    $e_1 = e_2 = e_3 = e_4 = (219)(0.25) = 54.75$, $\alpha = 0.05$, $\chi^2_{3, 0.05} = 7.81$.

    H_0: $P_1 = P_2 = P_3 = P_4 = 0.25$ vs. H_1: At least one P_i is different.

    TS: $\chi^2 = 2.571$.

    Conclusion: Since $2.571 < 7.81$, do not reject H_0. That is, there is insufficient evidence to claim that the proportions are different from those specified in the null hypothesis, at the 5% level of significance.

3. Answers will vary.

5. Below is a *Minitab* output with the relevant computations for this problem.

    H_0: The weight of the car and the severity of the injury are independent.

    H_1: The weight of the car and the severity of the injury are not independent.

    TS: $\chi^2 = 15.38$ with a p-value $= 0.004$.

    Conclusion: Since p-value $= 0.004 < \alpha = 0.05$, reject H_0. That is, there is sufficient sample evidence to conclude that the weight of the car and the severity of the injury are not independent.

```
┌─────────────────────────────────────────────────────────────────────┐
│                                                                       │
│   Chi-Square Test: Less than 2500, 2500 to 3000, Greater than 3000    │
│                                                                       │
│   Expected counts are printed below observed counts                   │
│   Chi-Square contributions are printed below expected counts          │
│                                                                       │
│                        Less      2500     Greater                     │
│                        than       to       than                       │
│                        2500      3000      3000      Total            │
│     Very Severe          34        22         8        64             │
│                        23.01     22.11     18.88                      │
│                        5.248     0.001     6.267                      │
│                                                                       │
│     Average              43        41        47       131             │
│                        47.10     45.26     38.64                      │
│                        0.357     0.401     1.810                      │
│                                                                       │
│     Moderate             51        60        50       161             │
│                        57.89     55.63     47.49                      │
│                        0.820     0.344     0.133                      │
│                                                                       │
│          Total         128       123       105       356             │
│                                                                       │
│   Chi-Sq = 15.380, DF = 4, P-Value = 0.004                            │
│                                                                       │
└─────────────────────────────────────────────────────────────────────┘
```

7. $n = 257$, $N_1 = 123$, $N_2 = 135$, $N_3 = 141$, $N_4 = 141$, $P_1 = P_2 = P_3 = P_4 = 0.25$,

$e_1 = e_2 = e_3 = e_4 = (527)(0.25) = 131.75$.

H_0: $P_1 = P_2 = P_3 = P_4 = 0.25$ vs. H_1: At least one P_i is different.

TS: $\chi^2 = 1.417$.

p-value $= P[\chi^2 > 1.417] > 0.5$ with 3 degrees of freedom

Conclusion: Since the p-value is rather large, do not reject H_0. That is, there is insufficient evidence to claim that the earthquakes are not equally likely to occur.

9. $n = 200$, $N_1 = 6$, $N_2 = 42$, $N_3 = 48$, $N_4 = 60$, $N_5 = 21$, $N_6 = 21$,

$P_1 = 0.0422$, $P_2 = 0.2484$, $P_3 = 0.2903$, $P_4 = 0.2728$, $P_5 = 0.0693$,

$P_6 = 0.077$, $e_1 = (200)(0.0422) = 8.44$, $e_2 = (200)(0.2484) = 49.68$,

$e_3 = (200)(0.2903) = 58.06$, $e_4 = (200)(0.2728) = 54.56$,

$e_5 = (200)(0..693) = 13.86$, $e_6 = (200)(0.077) = 15.4$, $\alpha = 0.05$,

$$\chi^2_{5, 0.05} = 11.07.$$

H_0: $P_1 = 0.0422$, $P_2 = 0.2484$, $P_3 = 0.2903$, $P_4 = 0.2728$, $P_5 = 0.0693$,

$P_6 = 0.077$

H_1: At least one P_i is different.

TS: $\chi^2 = 11.60697$.

Conclusion: Since $11.60697 < 11.07$, reject H_0. That is, there is sufficient sample evidence to claim that the height distribution has changed at the 5% level of significance.

11. Below is a *Minitab* output with the relevant computations for this problem.

H_0: Receiving the vaccine and not contracting the flu are independent.

H_1: Receiving the vaccine and not contracting the flu are not independent.

TS: $\chi^2 = 0.564$ with a p-value $= 0.453$.

Conclusion: Since p-value = 0.453 > α = 0.05, do not reject H_0. That is, there is insufficient sample evidence to conclude that receiving the vaccine and not contracting the flu are not independent at the 5% level of significance.

Same conclusion at the 1% and 10% significance levels.

```
Chi-Square Test: Vaccine, No Vaccine

Expected counts are printed below observed counts
Chi-Square contributions are printed below expected counts

                Vaccine       No Vaccine       Total
    Flu            10              6             16
                11.32           4.68
                0.155           0.374

  No Flu          174             70            244
               172.68          71.32
                0.010           0.025


    Total         184             76            260

Chi-Sq = 0.564, DF = 1, P-Value = 0.453
1 cells with expected counts less than 5.
```

13. Below is a *Minitab* output with the relevant computations for this problem.

   (a) H_0: The birth weight of the baby and the age of the mother are independent.

   H_1: The birth weight of the baby and the age of the mother are not independent.

   TS: $\chi^2 = 1.586$ with a p-value = 0.208.

Conclusion: Since p-value = 0.208 > α = 0.05, do not reject H$_0$. That is, there is insufficient sample evidence to conclude that birth weight of the baby and the age of the mother are not independent at the 5% level of significance.

(b) p-value = 0.208.

```
Chi-Square Test: Less Than 2500, More Than 2500

Expected counts are printed below observed counts
Chi-Square contributions are printed below expected counts

                        Less          More
                        Than          Than
                        2500          2500        Total
    20 or Less           12            50            62
                         9.07         52.93
                         0.944         0.162

    Greater than 20      18           125           143
                        20.93        122.07
                         0.409         0.070

            Total        30           175           205

Chi-Sq = 1.586, DF = 1, P-Value = 0.208
```

15. Below is a *Minitab* output with the relevant computations for this problem.

H$_0$: The course rating and whether the course is required are independent.

H$_1$: The course rating and whether the course is required are not independent.

TS: $\chi^2 = 1.862$ with a p-value = 0.395.

Conclusion: Since p-value = 0.395 > α = 0.05, do not reject H₀. That is, there is insufficient sample evidence to conclude that the course rating and whether the course is required are not independent at the 5% level of significance.

Same conclusion at the 1% level of significance.

```
Chi-Square Test: Excellent, Average, Poor

Expected counts are printed below observed counts
Chi-Square contributions are printed below expected counts

                    Excellent    Average     Poor     Total
     Required              14         42       18        74
                        15.90      42.81    15.29
                        0.227      0.015    0.481

 Not Required              12         28        7        47
                        10.10      27.19     9.71
                        0.358      0.024    0.757

        Total             26         70       25       121

Chi-Sq = 1.862, DF = 2, P-Value = 0.394
```

17. Observed data values will vary.

Chapter 14 NONPARAMETRIC HYPOTHESES TESTS

SECTION 14.2 - SIGN TEST

PROBLEMS

1. (a) Below is a partial *Minitab* output for this problem.

```
Sign Test for Median: BloodPressure

Sign test of median =   128.0 versus not = 128.0

                     N    Below  Equal  Above      P
BloodPressure      100      40      0     60   0.0574
```

H_0: $\eta = 128$ against H_1: $\eta \neq 128$.

TS: $N = 40$.

p-value = 2 Minimum[$P\{N \geq 40\}$, $P\{N \leq 40\}$] = 0.0574.
(using the normal approximation)

Conclusion: Since p-value = 0.0574 is rather large, do not reject the null hypothesis and conclude that the median systolic blood pressure of middle-aged men has not changed.

(b) Below is a partial *Minitab* output for this problem.

```
Sign Test for Median: BloodPressure

Sign test of median =   128.0 versus not = 128.0

                   N     Below   Equal   Above     P
BloodPressure     100      30       0      70    0.0001
```

H_0: $\eta = 128$ against H_1: $\eta \neq 128$.

TS: N = 30.

p-value = 2 Minimum[P{N $\geq$ 30}, P{N $\leq$ 30}] = 0.0001.
           (using the normal approximation)

Conclusion: Since p-value = 0.0001 is very small, reject the null hypothesis and conclude that the median systolic blood pressure of middle-aged men has changed.

(c) Below is a partial *Minitab* output for this problem.

```
Sign Test for Median: BloodPressure

Sign test of median =   128.0 versus not = 128.0

                   N    Below  Equal  Above      P
BloodPressure     100    20      0     80     0.0000
```

H_0: $\eta = 128$ against H_1: $\eta \neq 128$.

TS: N = 20.

p-value = 2 Minimum[P{N $\geq$ 20}, P{N $\leq$ 20}] = 0.0000.
(using the normal approximation)

Conclusion: Since p-value = 0.0000 is very small, reject the null hypothesis and conclude that the median systolic blood pressure of middle-aged men has changed.

3. Below is a partial *Minitab* output for this problem.

```
Sign Test for Median: Scores

Sign test of median =   0.00000 versus not = 0.00000

           N     Below   Equal   Above      P
Scores    50      21       0      29      0.3222
```

H_0: $\eta = 0$ against H_1: $\eta \neq 0$.

TS: N = 21.

p-value = 2 Minimum[P{N $\geq$ 21}, P{N $\leq$ 21}] = 0.3222.
(using the normal approximation)

Conclusion: For a significance level of 5%, since p-value = 0.3222 > 0.05, do not reject the null hypothesis. That is, there is insufficient sample evidence to claim that the two guns are not equally effective.

5. Below is a partial *Minitab* output for this problem.

```
Sign Test for Median: TestScores

Sign test of median =   72.00 versus > 72.00

              N     Below  Equal  Above      P     Median
TestScores   13       8      0      5     0.8666    69.00
```

H_0: $\eta \le 72$ against H_1: $\eta > 72$.

TS: $N = 8$.

p-value = $P\{N \le 8\}] = 0.8666$ (using the normal
approximation)

Conclusion: For a significance level of 5%, since
p-value = $0.8666 > 0.05$, do not reject the null hypothesis. That is,
there is insufficient sample evidence to claim that the median
score is at least 72.

7. Below is a partial *Minitab* output with some of the computations for this
problem.

```
Sign Test for Median: Weight

Sign test of median =   110.0 versus < 110.0

              N    Below   Equal  Above        P
Weight      200     120       0     80    0.0029
```

H_0: $\eta \ge 110$ against H_1: $\eta < 110$.

TS: $N = 120$.

p-value = $P\{N > 120\} = 0.0029$ (using the normal
approximation).

Conclusion: For a significance level of 5%, since
p-value = $0.0029 < 0.05$, reject the null hypothesis. That is,
there is sufficient sample evidence to claim that the median
weight of 16-year-old females from Los Angeles is less than
110 pounds.

SECTION 14.3 - SIGNED RANK TEST

PROBLEMS

1. (a) The table below shows the differences (DIF), the absolute value of the
   differences (ABSDIF), and the ranks of the absolute differences.

DIF	ABSDIF	RANKS
-17	17	5
33	33	8
22	22	7
-8	8	3
55	55	12
-41	41	11
-18	18	6
40	40	10
39	39	9
14	14	4
-88	88	13
99	99	14
102	102	15
-5	5	1
7	7	2

Since the test statistic (TS) is the sum of the ranks of the negative differences, then

$$TS = 5 + 3 + 11 + 6 + 13 + 1 = 39.$$

(b) The table below shows the differences (DIF), the absolute value of
The differences (ABSDIF), and the ranks of the absolute
differences. Since the test statistic (TS) is the sum of the ranks of
the negative differences, then

$$TS = 2 + 7 + 10 + 6 + 8 + 9 = 42.$$

DIF	ABSDIF	RANKS
44.00	44.00	11.5
2.00	2.00	5.0
1.00	1.00	3.0
-0.40	0.40	2.0
-3.00	3.00	7.0
-13.00	13.00	10.0
44.00	44.00	11.5
50.00	50.00	13.0
1.10	1.10	4.0
-2.20	2.20	6.0
0.01	0.01	1.0
-4.00	4.00	8.0
-6.60	6.60	9.0

(c) The table below shows the differences (DIF), the absolute value of the differences (ABSDIF), and the ranks of the absolute differences.

DIF	ABSDIF	RANKS
12	12	4
15	15	5
19	19	6
8	8	3
-3	3	1
-7	7	2
-22	22	7
-55	55	10
48	48	9
31	31	8
89	89	11
92	92	12

Since the test statistic (TS) is the sum of the ranks of the negative differences, then

$$TS = 1 + 2 + 7 + 10 = 20.$$

3. (a) $E[TS] = (15)(16)/4 = 60;$ $Var(TS) = (15)(16)(31)/24 = 310.$

   p-value $= 2 \, Min(P\{TS \geq 39\}, P\{TS \leq 39\}) = 2 \, P\{TS \leq 39.5\}$

   $\approx 2 \, P\{Z \leq (39.5 - 60)/\sqrt{310}\} = 2 \, P\{Z \leq -1.16\}$

   $= 2(1 - 0.8770) = 0.246.$

   (b) $E[TS] = (13)(14)/4 = 45.5;$ $Var(TS) = (13)(14)(27)/24 = 204.75.$

   p-value $= 2 \, Min(P\{TS \geq 42\}, P\{TS \leq 42\}) = 2 \, P\{TS \leq 42.5\}$

   $\approx 2P\{Z \leq (42.5 - 45.5)/\sqrt{204.75}\} = 2P\{Z \leq -0.21\}$

   $= 2(1 - 0.5832) = 0.8336.$

   (c) $E[TS] = (12)(13)/4 = 39;$ $Var(TS) = (12)(13)(25)/24 = 162.5.$

   p-value $= 2 \, Min(P\{TS \geq 20\}, P\{TS \leq 20\}) = 2 \, P\{TS \leq 20.5\}$

   $\approx 2P\{Z \leq (20.5 - 39)/\sqrt{162.5}\} = 2P\{Z \leq -1.45\}$

   $= 2(1 - 0.9265) = 0.147.$

5. (a) and (b) - The table below gives the differences and the rank of the absolute differences.

   $n = 14$, $\mu = 52.5$, $\sigma = 15.9295.$

   H_0: The two population distributions are equal.

   H_1: The two population distributions are not equal.

   TS: TS = 84.

p-value = 2P{TS ≥ 84} = 2P{TS ≥ 84.5}

= 2P{Z ≥ 2.01} = 0.0348.

Conclusion: For a significance level of 5%, since
p-value = 0.0348 < 0.05, reject the null hypothesis. That is,
there is sufficient sample evidence to claim that the way the paper
was presented (handwritten or typed) had an effect on the given score.

HAND	TYPED	DIF	ABS	RANK
83	88	-5	5	7.0
75	91	-16	16	14.0
75	72	3	3	2.5
60	70	-10	10	11.5
72	80	-8	8	10.0
55	65	-10	10	11.5
94	90	4	4	5.0
85	89	-4	4	5.0
78	85	-7	7	9.0
96	93	3	3	2.5
80	86	-6	6	8.0
75	79	-4	4	5.0
66	64	2	2	1.0
55	68	-13	13	13.0

7. (a) Below is a partial *Minitab* output with some of the computations for this problem.

```
Sign Test for Median: DIFFERENCE

Sign test of median =   0.00000 versus not = 0.00000

               N    Below  Equal  Above       P    Median
DIFFERENCE     8      6      1      1     0.1250   -30.00
```

H_0: $\eta = 0$ against H_1: $\eta \neq 0$.

TS: $N \doteq 6$.

p-value = 2 Minimum$[P\{N \geq 6\}, P\{N \leq 6\}] = 0.0625$.
(using the normal approximation)

Conclusion: For a significance level of 5%, since the p-value = 0.0625 > 0.05, do not reject the null hypothesis. That is, there is insufficient sample evidence to claim that the quoted price for repair depends on the gender of the person.

(b) The table below gives the differences and the rank of the absolute differences.

MAN	WOMAN	FIFFERENCE	ABS	RANK
145	145	0	0	1
220	300	-80	80	7
150	200	-50	50	6
100	125	-25	25	4
250	400	-150	150	8
150	135	15	15	2
180	200	-20	20	3
240	275	-35	35	5

$n = 7$, $\mu = 14$, $\sigma = 5.9161$.

H_0: The two population distributions are equal.

H_1: The two population distributions are not equal.

TS: TS = 33.

p-value = $2P\{TS \geq 33\} = 2P\{TS \geq 33.5\}$

$\qquad = 2P\{Z \geq 2.17\} = 0.015$.

Conclusion: Since the p-value = 0.015 is rather small, reject the null hypothesis. That is, there is sufficient sample evidence to claim that the quoted price for repair depends on the gender of the person.

9. The table below gives the differences and the rank of the absolute differences.

NOPAINT	PAINT	DIF	ABS	RANK
426.1	416.7	9.4	9.4	9
438.5	431.0	7.5	7.5	8
440.6	442.6	-2.0	2.0	3
418.5	423.6	-5.1	5.1	5
441.2	447.5	-6.3	6.3	6
427.5	423.9	3.6	3.6	4
412.2	412.8	-0.6	0.6	1
421.0	419.8	1.2	1.2	2
434.7	424.1	10.6	10.6	10
411.9	418.7	-6.8	6.8	7

$n = 10$, $\mu = 27.5$, $\sigma = 9.811$.

H_0: The two population distributions are equal.

H_1: The two population distributions are not equal.

TS: TS = 22.

p-value $= 2P\{TS \le 22\} = 2P\{TS \le 22.5\}$

$$= 2P\{Z \le 0.51\} = 0.305.$$

Conclusion: For a significance level of 5%, since the
p-value $= 0.305 > 0.05$, do not reject the null hypothesis. That is, there
is insufficient sample evidence to claim that the cruising speed of the
aircraft depends on whether the craft is painted or not.

SECTION 14.4 - RANK-SUM TEST FOR COMPARING TWO POPULATIONS

PROBLEMS

1. Below is a listing of the sample values and their corresponding ranks.

Sample 1	142	155	237	244	202	111	326	334	350	247
Rank	3	4	9	10	7	1	15	16	18	11
Sample 2	212	277	175	138	341	255	303	188		
Rank	8	13	5	2	17	12	14	6		

(a) The sum of the ranks for Sample 1: $3 + 4 + ... + 18 + 11 = 94$.
(b) The sum of the ranks for Sample 2: $8 + 13 + ... + 14 + 6 = 77$.

3. The samples and their ranks are given in the table below.

Rural	544	567	475	658	590	602	571	502	578
Rank	7	9	1	17	12	13	10	3	11
Urban	610	498	505	711	545	613	509	514	609
Rank	15	2	4	18	8	16	5	6	14

Let n be the sample size for the rural group and m for the urban group. We will use the rural group to compute the TS.

$n = m = 9$. Sum of ranks for the rural group = 83. E[TS]=85.5.

Var(TS) = 128.25.

H_0: Distributions of SAT scores in both counties are the same.

H_1: Distributions of SAT scores in both counties are not the same.

TS:W=83.

p-value = $2P\{W \le 83\} = 2P\{W \le 83.5\} \approx 2P\{Z \le -0.18\} = 0.8572$.

Conclusion: At the 5% level of significance, since the
p-value = $0.8572 > 0.05$, do not reject the null hypothesis. That is, there is
insufficient sample evidence to conclude that distributions of SAT scores
in both counties are not the same.

5. The samples and their ranks are given in the table below.

Stanford	57.8	60.4	71.2	52.5	68	69.6	70	54	48.8	57.6
Rank	12	13	20	4	17	18	19	6	2	11
Berkeley	52.6	56.6	61	47.9	55	62.5	66.4	57.5	56.5	49.8
Rank	5	9	14	1	7	15	16	10	8	3

Let n be the sample size for the Stanford group and m for the Berkeley group. We will use the Stanford group to compute the TS.

$n = m = 10$. Sum of ranks for the Untreated group =122. E[TS] =105.

Var(TS)=175.

H_0: Starting salary distribution for MBA graduates from Stanford and
    Berkeley are the same.

H_1: Starting salary distribution for MBA graduates from Stanford
    and Berkeley are not the same.

TS: W = 122.

p-value = $2P\{W \ge 122\} = 2P\{W \ge 121.5\} \approx 2P\{Z \ge 1.25\} = 0.2112$.

Conclusion: At the 5% level of significance, since the
p-value = 0.2112 > 0.05, do not reject the null hypothesis. That is, there
is insufficient sample evidence to conclude that starting salary
distribution for MA graduates from Stanford and Berkeley are not the
same.

7. $n = m = 12$, TS = 336, E[TS] = 150, Var(TS) = 300.

p-value = $2P\{W \geq 336\} = 2P\{W \geq 335.5\} \approx 2P\{Z \geq 10.70\} = 0.0000$.

9. p-value = 0.2123.

SECTION 14.5 RUNS TEST FOR RANDOMNESS

PROBLEMS
1. (a) Largest possible number of runs = 41.

   (b) Smallest possible number of runs = 2.

3. Total number of runs R = 7.

   Let n be the number of acceptable watches and m the number of unacceptable watches.

   $n = 19$, $m = 7$, $E[TS] = 11.2308$, $Var(TS) = 3.7775$.

   H_0: Number of acceptable watches in the production run is random.

   H_1: Number of acceptable watches in the production run is not random.

   TS: R = 7.

   p-value = $2P\{R \le 7\} = 2P\{R \le 7.5\} \approx 2P\{Z \le -1.92\} = 0.0528$.

   Conclusion: Since the p-value = 0.0528 > 0.05, do not reject the null hypothesis. That is, there is insufficient sample evidence to conclude that the number of acceptable watches in the production run is not random.

5. Let n be the number of females and m the number of males.

   $n = 10$, $m = 15$, $R = 8$, $E[TS] = 13$, $Var(TS) = 5.5$.

   H_0: The chosen order was random.

   H_1: The chosen order was not random.

   TS: R = 8.

p-value $= 2P\{R \le 8\} = 2P\{R \le 8.5\} \approx 2P\{Z \le -1.9188\}$

$$= 2(0.0274) = 0.0548.$$

Conclusion: Since the p-value $= 0.0548$ is rather large, do not reject the null hypothesis. That is, there is insufficient sample evidence to conclude that the chosen order for the interview was not random.

7. Below is a partial *Minitab* output for the problem.

```
Descriptive Statistics: Hours

Variable    N   N*    Mean  SE Mean  StDev  Variance  Minimum      Q1  Median
Hours      30    0  165.40    2.57  14.10    198.73   138.00  153.75  163.50

Variable     Q3  Maximum    IQR
Hours    177.25   191.00  23.50


Runs Test: Hours

Runs test for Hours

Runs above and below K = 163.5

The observed number of runs = 8
The expected number of runs = 16
15 observations above K, 15 below
P-value = 0.003
```

(a) The sample median $S_m = 163.5$.

(b) The number of runs, $R = 8$.

(c) Let n be the number of values below the sample median and let m be the number above the sample median.

$n = m = 15$, $E[TS] = 16$, $Var(TS) = 7.2414$.

H_0: The sequence of values constitutes a random sample.

H_1: The sequence of values does not constitute a random sample.

TS: $R = 8$.

p-value $= 2P\{R \le 7\} = 2P\{R \le 7.5\} \approx 2P\{Z \le -3.1587\} = 0.0016$.

Note: The p-value in the output is 0.003, which is much more accurate since the number of decimal places used by the computer was more than four.

Conclusion: Since the p-value $= 0.0016$ is rather small, reject the null hypothesis. That is, there is sufficient sample evidence to conclude that the sequence of values does not constitute a random sample

Note: Small sample size may make the approximation invalid.

REVIEW PROBLEMS

1. The samples and their ranks are given in the table below.

July	32.2	27.4	28.6	32.4	40.5	26.2	29.4	25.8	36.6	30.3	28.5	32
Rank	13	3	6	14	23	2	7	1	17	10	5	12
Jan.	30.5	28.4	40.2	37.6	36.5	38.8	34.7	29.5	29.7	37.2	41.5	37
Rank	11	4	22	20	16	21	15	8	9	19	24	18

Let n be the sample size for July and m for January. We will use the July sample to compute the TS.

n = m = 12. Sum of ranks for July = 113.

E[TS] = 150, Var(TS) = 300.

H_0: The distribution of caloric intake by females in January and July are the same.

H_1: The distribution of caloric intake by females in January and July are not the same.

TS: W = 113.

p-value = $2P\{W \le 113\} = 2P\{W \le 113.5\} \approx 2P\{Z \le -2.11\} = 0.0348$.

Conclusion: At the 5% level of significance, since the p-value = 0.0348 < 0.05, reject the null hypothesis. That is, there is insufficient sample evidence to conclude that the distribution of caloric intake by females in January and July are not the same.

3. Below is a partial *Minitab* output with some of the computations for this problem.

Sign Test for Median: NetWorth

```
Sign test of median =   104.5 versus  <  104.5

                N      Below  Equal   Above        P
NetWorth      1000      579      0     421     0.0000
```

H_0: $\eta \geq 104.5$ against H_1: $\eta < 104.5$.

TS: $N = 579$.

p-value $= P\{N \geq 579\} = 0.0000$ (using the normal approximation).

Conclusion: For a significance level of 5%, since
p-value $= 0.0000 < 0.05$, reject the null hypothesis. That is,
there is sufficient sample evidence to claim that the median net worth has
decreased.

5. Below is a partial *Minitab* output for this problem.

   The sample median $S_m = 145$. The number of runs, $R = 21$.

   Let n be the number of values below the sample median and let m be
   the number above the sample median.

   $n = m = 20$, $E[TS] = 21$, $Var(TS) = 9.7436$.

   H_0: The sequence of weights constitutes a random sample.

   H_1: The sequence of weights does not constitute a random sample.

   TS: $R = 21$.

   p-value $= 2P\{R \geq 21\} = 2P\{R \geq 21.5\} \approx 2P\{Z \geq 0.16\} = 0.8728$.

Conclusion: Since the p-value = 0.8728 is very large, do not reject the null hypothesis. That is, there is insufficient sample evidence to conclude that the sequence of weights does not constitute a random sample.

```
Descriptive Statistics: Weight

Variable   N  N*    Mean  SE Mean  StDev  Minimum      Q1  Median      Q3
Weight    40   0  149.38     5.44  34.39   102.00  126.25  145.00  167.50

Variable  Maximum
Weight     249.00


Runs Test: Weight

Runs test for Weight

Runs above and below K = 145

The observed number of runs = 21
The expected number of runs = 21
20 observations above K, 20 below
P-value = 1.000
```

7. Answers will vary.

9. The table below gives the differences and the rank of the differences.

WEEK1	WEEK2	DIF	RANK
46	54	-8	4
54	60	-6	5
74	96	-22	1
60	75	-15	3
63	80	-17	2
45	50	-5	6

$n = 6$, $\mu = 10.5$, $\sigma = 4.7697$.

H_0: The two population distributions are equal.

H_1: The two population distributions are not equal.

TS: TS = 21.

p-value = $2P\{TS \geq 21\} = 2P\{TS \geq 20.5\}$

$= 2P\{Z \geq 2.0966\} = 0.0444$.

Conclusion: Since the p-value = 0.0444 (< 0.05) is rather small, reject the null hypothesis. That is, there is sufficient sample evidence to claim that the two population distributions are not equal. There seems to be an increase in sales after the advertisement campaign since all the differences (WEEK1 - WEEK2) are negative.

11. The samples and their ranks are given in the table below.

Man	145	220	150	100	250	150	180	240
Rank	4.5	11	6.5	1	13	6.5	8	12
Woman	145	300	200	125	400	135	200	275
Rank	4.5	15	9.5	2	16	3	9.5	14

Let n be the sample size for the male group and m for the female group. We will use the male group to compute the TS.

n = m = 8. Sum of ranks for the rural group = 62.5. E[TS] = 68.

Var(TS) = 90.6667.

H_0: Distributions of price quotes received by the man and woman are the same.

H_1: Distributions of price quotes received by the man and woman are not the same.

TS:W = 62.5.

p-value = 2P{W ≤ 62.5} ≈ 2P{Z ≤ -0.5776} = 0.5754.

Conclusion: At the 5% level of significance, since the p-value = 0.5754 > 0.05, do not reject the null hypothesis. That is, there is insufficient sample evidence to conclude that the distribution of price quotes received by the man and woman are not the same.

Chapter 15 QUALITY CONTROL

SECTION 15.2   THE X-BAR CONTROL CHART FOR DETECTING A SHIFT IN THE MEAN

PROBLEMS

1. $\mu = 100, \sigma = 10$.

   (a) $n = 4$.

   $$LCL = \mu - 3\sigma/\sqrt{n} = 85 \text{ and } UCL = \mu + 3\sigma/\sqrt{n} = 115.$$

   (b) $n = 5$.

   $$LCL = \mu - 3\sigma/\sqrt{n} = 86.5836 \text{ and } UCL = \mu - 3\sigma/\sqrt{n} = 113.4164.$$

   (c) $n = 6$.

   $$LCL = \mu - 3\sigma/\sqrt{n} = 87.7526 \text{ and } UCL = \mu + 3\sigma/\sqrt{n} = 112.2474.$$

   (d) $n = 10$.

   $$LCL = \mu - 3\sigma/\sqrt{n} = 90.5132 \text{ and } UCL = \mu + 3\sigma/\sqrt{n} = 109.4868.$$

3. $n = 5, \mu = 80, \sigma = 10$.

   $$LCL = \mu - 3\sigma/\sqrt{n} = 66.5836 \text{ and } UCL = \mu + 3\sigma/\sqrt{n} = 93.4164.$$

   The process is out of control since the average for the 9th subgroup of 94 > 93.4164 (UCL). The following x-bar graph depicts this situation.

Chapter 15   Quality Control

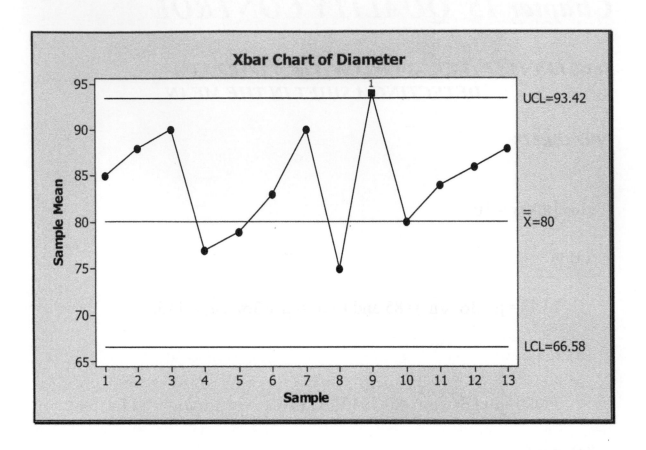

5. $n = 5$, $\mu = 0$, $\sigma = 0.005$.

LCL $= \mu - 3\sigma/\sqrt{n} = -0.0067$ and UCL $= \mu - 3\sigma/\sqrt{n} = 0.0067$.

The process is in control since none of the averages fall outside the control limits. This situation is depicted in the following graph.

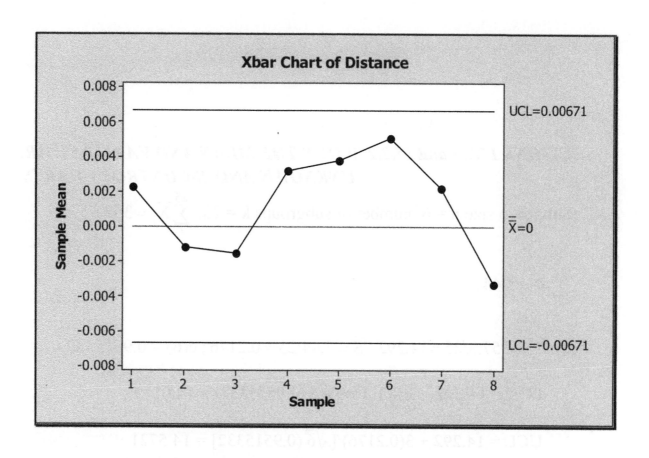

SECTIONS 15.2.1 and 15.2.2   *WHEN THE MEAN AND VARIANCE ARE UNKNOWN AND S-CONTROL CHARTS*

1. Subgroup size n = 6; number of subgroups k = 25; $\sum_{i}^{25} \overline{X}_i = 357.3$;

   $\sum_{i}^{25} S_i = 5.44$.

   (a) $\overline{\overline{X}} = 357.3/25 = 14.292$; $\overline{S} = 5.44/25 = 0.2176$; c(6) = 0.9515332.

   LCL = 14.292 – 3(0.2176)/[$\sqrt{6}$ (0.9515332] = 14.0119

   UCL = 14.292 + 3(0.2176)/[$\sqrt{6}$ (0.9515332] = 14.5721

   (b) Percentage = (14.5721 – 14.0119)/(14.8 – 13.8)]×100% = 56.02%.

3. (a) $\overline{\overline{X}} = 15.88$, $\overline{S} = 2.80$.

   (b) Following are the x-bar and S-charts for the data. One can observe from the S-chart that the process standard deviation is out of control.

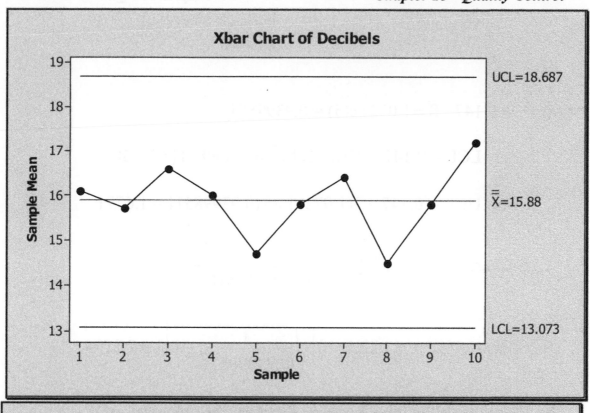

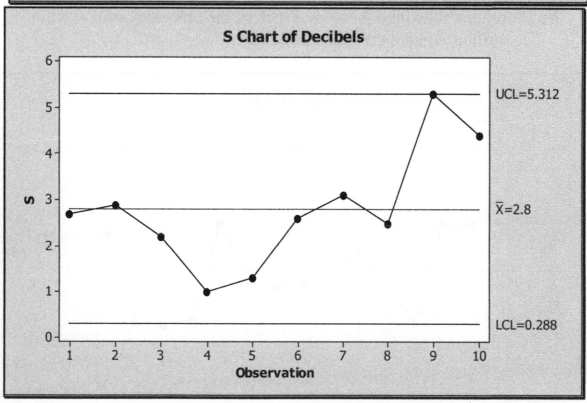

5. (a) $\overline{\overline{X}} = 0.147$; $\overline{S} = 1.072$; $c(5) = 0.9399851$.

x-Bar: LCL $= 0.147 - 3(1.072)/[\sqrt{5}\,(0.9399851] = -1.3831$

x-Bar: UCL $= 0.147 + 3(1.072)/[\sqrt{5}\,(0.9399851] = 1.6771$

S-Chart: LCL $= 1.072\left[1 - 3\left[\sqrt{\dfrac{1}{(0.9399851)^2} - 1}\,\right]\right] = -0.0954$

S-Chart: UCL $= 1.072\left[1 + 3\left[\sqrt{\dfrac{1}{(0.9399851)^2} - 1}\,\right]\right] = 2.2394$

(b) From the following x-bar- and S-charts, the process seem to be in control throughout the observed values.

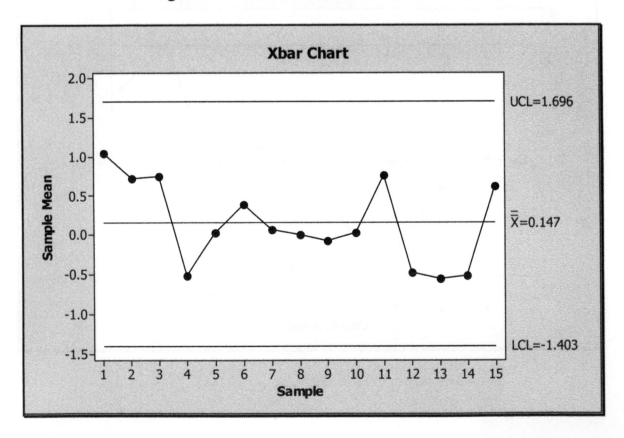

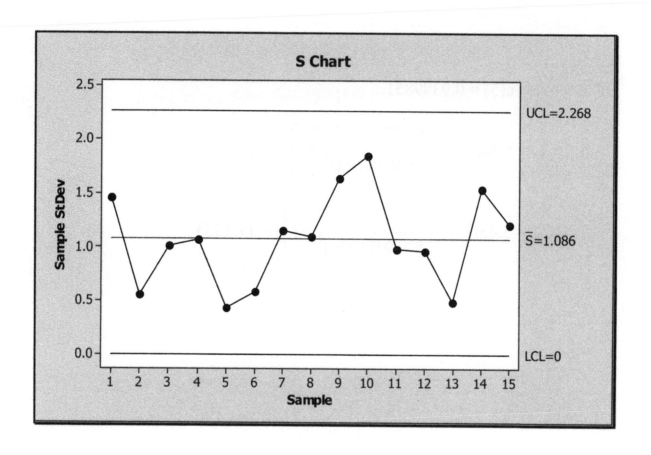

7. $\overline{\overline{X}} = 36.022$; $\overline{S} = 4.294$; $c(5) = 0.9399851$.

x-Bar: LCL $= 36.022 - 3(4.294)/[\sqrt{5}\,(0.9399851)] = 29.8931$

x-Bar: UCL $= 36.022 + 3(4.294)/[\sqrt{5}\,(0.9399851)] = 42.1508$

S: LCL $= 4.294[1 - 3\left[\sqrt{\dfrac{1}{(0.9399851)^2} - 1}\right]] = -0.3822$

S: UCL $= 4.294[1 + 3\left[\sqrt{\dfrac{1}{(0.9399851)^2} - 1}\right]] = 8.9702$

9. $\overline{S} = 5.43$, c(5) = 0.9399851.

$$S: LCL = 5.43\left[1 - 3\left[\sqrt{\frac{1}{(0.9399851)^2}} - 1\right]\right] = -0.4833$$

$$S: UCL = 5.43\left[1 + 3\left[\sqrt{\frac{1}{(0.9399851)^2}} - 1\right]\right] = 11.3433$$

SECTION 15.3   CONTROL CHARTS FOR FRACTION DEFECTIVE

PROBLEMS

1. n = 200, p = 0.03.

    LCL = np - 3 $\sqrt{np(1-p)}$ = -1.2374 or 0.0.

    UCL = np + 3 $\sqrt{np(1-p)}$ = 13.2374.

    The process is in control since none of the number of defectives fall outside the control limits.

3. (a) and (b) n = 100, p = 0.04.

    LCL = np - 3 $\sqrt{np(1-p)}$ = -1.8788 or 0.0.

    UCL = np + 3 $\sqrt{np(1-p)}$ = 9.8788.

    The process is in control since none of the number of defectives fall outside the control limits.

SECTION 15.4   EXPONENTIALLY WEIGHTED MOVING-AVERAGE CONTROL CHARTS

PROBLEMS

1. (a) The following displays the x-bar chart for the given data. Observe, based on this x-bar chart, the process is in control.

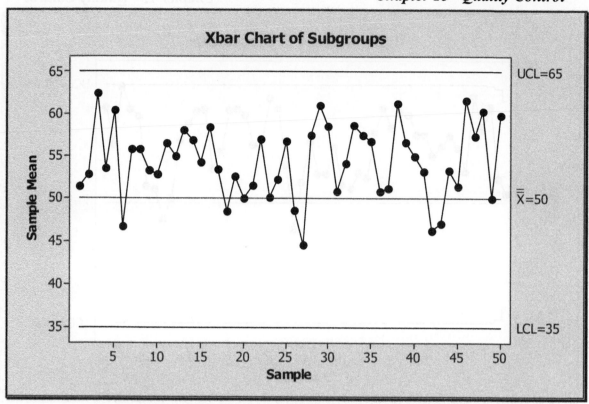

Xbar Chart of Subgroups

(b) The following EMWA chart with β = 0.5 will indicate that the process is in control.

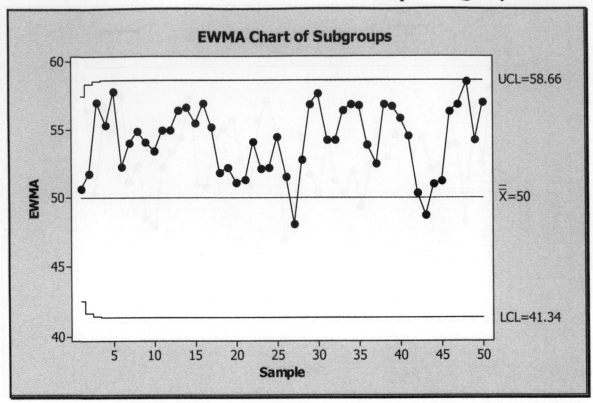

(c) The following EMWA chart with β = 0.25 will detect that the process is out of control.

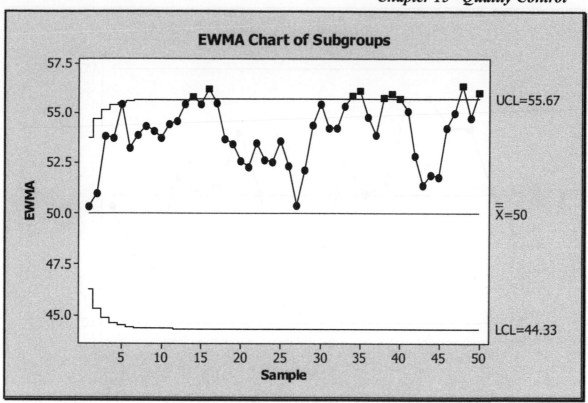

EWMA Chart of Subgroups

3. (a) The following EMWA chart with β = 0.7 will detect that the process is in control.

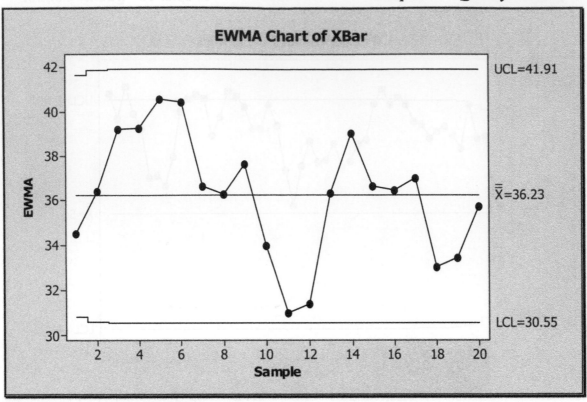

EWMA Chart of XBar

UCL=41.91

$\overline{\overline{X}}$=36.23

LCL=30.55

(b) Yes. The process was in control throughout.

(c) $\overline{\overline{X}} = 36.23$; $\overline{S} = 5.43$; $c(5) = 0.9399851$, $\overline{S}/c(5) = 5.7767$.

Hence, the SD(EWMA) = $\sqrt{\dfrac{0.7}{2-0.7}} \times \dfrac{5.7767}{\sqrt{5}} = 1.8957$. Thus,

$P\{25 \leq X \leq 45\} = P\{(25 - 36.230)/1.8957 \leq Z \leq (45 - 36.230)/1.8957\} =$

$P\{-5.9239 \leq Z \leq 4.6263\} \approx 1.$

SECTION 15.5   CUMULATIVE SUM CONTROL CHARTS

PROBLEMS

1. Problem does not exist.

3. $\overline{\overline{X}} = 15.88$, $\overline{S} = 2.8$; $d = 0.5$, $B = 4.77$, control limit $= \dfrac{B\sigma}{\sqrt{n}} =$

$\dfrac{4.77 \times 2.8 / 0.9399851}{\sqrt{4}} = 7.1044$; $Y_j = \overline{X}_j - \mu - \dfrac{d\sigma}{\sqrt{n}} = \overline{X}_j - 16.6247$.

$Y_1 = -0.5247$, $Y_2 = -0.9247$, $Y_3 = -0.0247$, $Y_4 = -0.6247$, $Y_5 = -1.9247$,

$Y_6 = -0.8247$, $Y_7 = -0.2247$, $Y_8 = -2.1247$, $Y_9 = -0.8247$, $Y_{10} = 0.5753$.

$S_1 = 0$, $S_2 = 0$, $S_3 = 0$, $S_4 = 0$, $S_5 = 0$, $S_6 = 0$, ..., $S_{10} = 0.5753$.

Since the control limit is $\dfrac{B\sigma}{\sqrt{n}} = 7.1044$, the cumulative-sum chart would not declare that the mean has increased after observing the first ten subgroup averages.

REVIEW PROBLEMS

1. $n = 4$, $\mu = 1.5$, $\sigma = 0.001$.

   $LCL = \mu - 3\sigma/\sqrt{n} = 1.4985$ and $UCL = \mu - 3\sigma/\sqrt{n} = 1.5015$.

3. $n = 300$, $p = 0.015$.

   $LCL = np - 3\sqrt{np(1-p)} = -1.8161$ (for practical purposes is equal to

   zero).

   $UCL = np + 3\sqrt{np(1-p)} = 10.8161$.

Printed and bound by CPI Group (UK) Ltd, Croydon, CR0 4YY

03/10/2024

01040315-0017